Lecture Notes in Control and Information Sciences 442

For further volumes:
http://www.springer.com/series/642

Series Editors [illegible]

[illegible]

[illegible]

P. Kokotović
University of California, Santa Barbara, CA, USA

A.B. Kurzhanski
Moscow State University, Russia

H. Kwakernaak
University of Twente, Enschede, The Netherlands

A. Rantzer
Lund Institute of Technology, Sweden

J.N. Tsitsiklis
MIT, Cambridge, MA, USA

For further volumes:
[illegible]

Xiaolian Zheng and Ben M. Chen

Stock Market Modeling and Forecasting

A System Adaptation Approach

Springer

Xiaolian Zheng
Department of Electrical and
Computer Engineering
National University of Singapore
Singapore

Ben M. Chen
Department of Electrical and
Computer Engineering
National University of Singapore
Singapore

ISSN 0170-8643
ISSN 1610-7411 (electronic)
ISBN 978-1-4471-5154-8
ISBN 978-1-4471-5155-5 (eBook)
DOI 10.1007/978-1-4471-5155-5
Springer London Heidelberg New York Dordrecht

Library of Congress Control Number: 2013936055

Printed on acid-free paper

Springer is part of Springer Science+Business Media (www.springer.com)

To My Parents
X. Zheng

To My Granduncle, Paul, and My Dad, Joseph
B. M. Chen

Preface

In recent decades, modeling of financial markets has aroused great interests in the academia and in the financial industry worldwide. Financial market is a highly complex system involving a large number of interacting factors ranging from psychological, social and political aspects to general economic performance. There is no conclusive knowledge of the influence of these factors and the corresponding market responses, even the public opinions on decisive factors to a particular financial event are usually diversified. The financial market is also known for its noisy data, nonlinearity and dynamic behavior, all of which make its modeling extremely difficult. These challenges call for a comprehensive study of the market behavior. Systems theory provides a totally different angle to investigate financial market modeling problems by analyzing the market as a complex system. It is a powerful tool in studying the interconnections and interactions inside the market and between the market and its environment. A framework that combines various foundations of both systems concepts and financial markets can thus provide more meaningful insights into market behaviors.

This monograph aims to develop a general framework or model structure based on systems theory to depict and analyze the financial markets, in particular the stock markets, and to provide accurate predictions of market prices and major market revisions, i.e., turning periods. The framework is carefully designed to consider both internal characters of the market and external influential factors, and thus is capable of capturing different types of market dynamics and behaviors. The structure of the framework and the processing of information inside and outside the market are to be addressed in detail. The applications of our framework in finding features in the stock markets from the U.S., China, Hong Kong, and Singapore, examples of both developed and emerging markets, are given and analyzed. Some interesting observations related to our framework and results, and possible future research directions that would further extend our study and understanding of the stock market, are also to be highlighted.

The intended audience for this monograph includes graduate students, researchers in areas related to financial modeling and analysts in financial sectors. It is assumed that the reader has some previous knowledge in financial modeling and statistics as well as in basic systems theory.

The authors have benefited a great deal from many fruitful discussions with our colleagues and research teammates at the National University of Singapore. Particularly, we would like to thank Dr. Nan Jiang, Dr. Feng Lin, Mr. Limiao Bai, Professor Delin Chu, Professor Qing-Guo Wang and Professor Cheng Xiang for their valuable comments and help during the course of studies and preparation of this monograph. We are also indebted to Dr. Sen Yan of Xiamen University, Ms. Jie Zheng of Rice University, Dr. Shanle Wu of UBS AG and Dr. Lichao Cheng of China Life Asset Management Company for their insightful suggestions and their generous assistance.

This monograph was typeset by the authors using LaTeX. All simulations, case studies and numerical computations were carried out in MATLAB. Diagrams were generated using XFIG and MATLAB with SIMULINK.

Kent Ridge, Singapore

Xiaolian Zheng
Ben M. Chen

Contents

Abbreviations

AIC	Akaike Information Criterion
ANN	Artificial Neural Network
AMH	Adaptive Market Hypothesis
AR	Autoregressive
ARCH	Autoregressive Conditional Heteroscedasticity
ARIMA	Autoregressive Integrated Moving Average
ARMA	Autoregressive Moving Average
ARMAX	Autoregressive Moving Average Model with Exogenous Input
BDI	Baltic Dry Index
CAPM	Capital Asset Pricing Model
CBOE	Chicago Board Options Exchange
CBR	Case Based Reasoning
CNY	Chinese Yuan
CPI	Consumer Price Index
DJIA	Dow Jones Industrial Average
DEFFR	Daily Effective Federal Funds Rate
EIU	Economist Intelligence Unit
EMA	Exponential Moving Average
EMH	Efficient Market Hypothesis
EUR/JPY	Euro versus Japanese Yen
FFFR	Federal Funds Future Rate
FFR	Federal Funds Rate
FFRT	Federal Funds Rate Target
FFT	Fast Fourier Transform
FMH	Fractal Market Hypothesis
FTSE	Financial Times and Stock Exchange
GARCH	Generalized Autoregressive Conditional Heteroskedasticity
GDP	Gross Domestic Product
GUI	Graphical User Interfaces
GUIDE	GUI Development Environment
HKAB	Hong Kong Association of Banks

HKD	Hong Kong Dollar
HKDISR	Hong Kong Dollar Interest Settlement Rates
HSI	Hang Seng Index
IFRI	Inflation Rate Indicator
IRI	Interest Rate Indicator
ISMI	International Stock Market Indicator
JPY	Japanese Yen
LIBOR	London Inter Bank Offered Rate
LM	Lagrange Multiplier
MA	Moving Average
MACD	Moving Average Convergence/Divergence
MAE	Mean Absolute Error
MARMA	Multivariate Autoregressive Moving Average
MAS	Monetary Authority of Singapore
MISO	Multiple Input and Single Output
ML	Maximum Likelihood
NASDAQ	National Association of Securities Dealers Automated Quotations
NVR	Noise Variance Ratio
NYSE	New York Stock Exchange
OE	Output Error
OP	Oil Price
PPI	Producer Price Index
RBF	Radial Basis Function
RWH	Random Walk Hypothesis
RMSE	Root Mean Squared Error
S&P	Standard & Poor's
SGD	Singapore Dollar
SGD/USD	Singapore Dollar versus U.S. Dollar
SGX	Singapore Stock Exchange
SHIBOR	Shanghai Interbank Offered Rate
SIBOR	Singapore Interbank Overnight Rate
SORA	Singapore Overnight Rate Average
SSE	Shanghai Stock Exchange
STI	Straits Times Index
STII	Straits Times Industrial Index
SVM	Support Vector Machine
T-TAS	Toolkit for Technical Analysis of Stocks
USD	U.S. Dollar
USD/CNY	U.S. Dollar versus Chinese Yuan
VAR	Vector Autoregressive
VHSI	HSI Volatility Index
VIX	Chicago Board Options Exchange Volatility Index
VXD	Chicago Board Options Exchange DJIA Volatility Index

Chapter 1
Introduction

1.1 Introduction

Stock, which is issued in the form of shares, is a certification of the ownership of a company. It is a type of security, "a legal representation of the right to receive prospective future benefits under stated conditions" [117]. A share of stock represents a unit of ownership. There are two classes of stock: the common stock and the preferred stock. Majority of the stock issued is usually referred to as the common stock. It is a residual claim but with voting rights. The holders of the preferred stock have a fixed or predetermined dividend and a superior priority over the common stockholders on the payment of dividend in the event of liquidation, but they usually do not have voting rights. When a bankruptcy occurs, the common stockholders will only be paid after the creditors, the bondholders and the preferred shareholders are paid off.

As it is an important and quick source to raise money, companies issue shares of stock. For example, if a company needs fresh capital to develop its business, it usually borrows money from a bank or sells part of the company as stock. The latter is called issuing shares of stock. Part of the ownership of the company will be transferred into shares of stock with its price being proportional to the value of the company. With the amount of money raised by selling these shares of stock, the company can then expand their business and enlarge their profits. If they succeed, the company will be valued more than before and the price of their company's stock will go up accordingly. This rapid development of business will bring stockholders more profits.

The stock market is a place for issuing, buying and selling shares of stock. Allured by the potential of significant profits that may be generated from it, much effort has been dedicated to the study of the stock market to better understand its influential factors, working mechanisms and market features. The stock market is a complex system involving various interacting factors from social, political to psychological aspects [45]. Financial markets function according to the basic economic theory of demand and supply. The market price of a stock is also determined by the interaction of aggregate demand and supply schedules [117]. Specifically, stock

X. Zheng & B.M. Chen: *Stock Market Modeling and Forecasting*, LNCIS 442, pp. 1–11.
DOI: 10.1007/978-1-4471-5155-5_1

prices are determined by fundamental elements in the long run, while psychological determinants or investors' consensus about the value of the company greatly affects short-term prices. As the economic condition is an important fundamental factor, stock prices react sensitively to economic news [29]. In other words, the stock market is considered as a mirror of the economy. Other important fundamental elements include the performance of a company, sector changes, management of a company, to name a few. All these impact the demand and supply of a stock in the long term. In the short run, investor sentiment which refers to the psychology of market participants directly affects the balance of demand and supply thus drives the stock prices. As pointed out by Graham [57], "in the short term, the stock market behaves like a voting machine". As a result, stock prices of a particular company may swing all the time, whereas its fundamental corporate value likely changes in a much slower pace.

The predictability of the stock market is one of the most important and attractive features for researchers and investors. With regard to this, there are two totally opposing opinions. Some people believe stock prices are unpredictable because the stock market is efficient and the price evolves in a random walk fashion; others believe that it is predictable, at least to a certain extend. The former believes the market is extremely efficient that prices always fully and instantaneously reflect available information. In this way, the price fluctuation is totally stochastic. This is explained in the famous efficient market hypothesis (EMH). The EMH was first proposed in 1900 by Bachelier [17, 18], who was also the first person to use the technique of Wiener process to model the behavior of stock prices. It was then further developed and refined in both theoretical and empirical aspects by many researchers including Fama [46, 47], Samuelson [112], and Roberts [109]. Among them, Fama [47] refined the EMH and defined the famous "strong, semistrong and weak" forms of market efficiency depending on the level of available information. The weak-form efficiency claims that all the historical public information has been fully reflected in prices. By supporting the weak-form efficiency, the semistrong-form efficiency additionally claims that prices will be instantly adjusted to reflect new public information. Lastly, in the strong-from counterpart, it is considered that all the information even the hidden or inside one is reflected in the prices, thus no excess profits can be gained besides of the price itself. This work triggered intensive investigations from the academic and industrial sectors. The EMH implies but is not equivalent to the random walk hypothesis (RWH), a theory that characterizes a price series as a random walk process. A renowned book, *A Random Walk Down Wall Street* [93], by Malkiel in 1973, has popularized the RWH theory and is still regarded as classic nowadays. Following the idea that the past information cannot be used to predict future prices, the RWH states that the changes of stock prices are independent of each other. In another words, the more efficient the market is, the more randomly the sequence of price changes. In many ways, the EMH and the random walks were proven to be different ideas, neither of them is necessary nor sufficient for each other [87, 92].

The EMH reached its ascendancy during the 1970s, but has suffered severe setbacks since then. Most arguments in favor of the EMH are supported by the statistical testing results, in which no predictive power of investigated models was shown and the normal distribution of price changes was presented. However, statistical evidences have largely been found to be contradictory to the EMH, such as leptokurtic, fat-tailed and negative skewed distribution of stock returns [1, 3, 6, 127], mean reversion [14, 48], and seasonality effects in stock returns [5, 21]. The arguments against the EMH always refer to the irrationality of the market, as one basic assumption of the EMH is that investors in the market are all rational. This has been supported by the emerging discipline of behavioral economics and behavioral finance. Besides this, the time delay always exists when the market reacts to new information. In this way, the instantaneous assimilation proposed by the EMH is considered to be unrealistic. Nowadays, the stock prices are believed to be the near-random-walk series, and thus have limited predictability. A classic investment book, *A Non-Random Walk Down Wall Street* [90] collected many statistical studies, proved that the stock market is predictable to some degree due to the inefficiency of the market.

There are two alternatives to the EMH, namely, the adaptive market hypothesis (AMH) and the fractal market hypothesis (FMH). The AMH explains the irrationality of the market as a rational reaction adapting to a changing environment. It views the market as an ecological system in which arbitrage opportunities exist and investment strategies will perform well in certain environments and poorly in others. By applying the principles of evolution to the financial interactions such as competition, mutation, reproduction, and natural selection, Lo [89] proposed a new framework that reconciled market efficiency with behavioral finance. The theoretical and empirical foundation for behavioral finance is built up by the book of Shleifer [118] as an alternative to the EMH. According to behavioral finance, markets are driven by psychological factors such as fear and greed. The ideas of Lo significantly attributed to the establishment of the AMH. From a deterministic perspective, the fractal market hypothesis (FMH) is another alternative to the EMH. It is proposed by Peters [103, 104] based on chaos theory. His work popularized the concept of chaos in the financial field. Stock prices exhibit the stochastic behavior, but it is believed that there are some deterministic features hidden behind. The complex properties of chaotic dynamics provide better explanations for this behavior. Therefore, chaos theory, an important part in dynamical systems (especially in nonlinear dynamics), aroused a great interest among researchers in the economic sector. However, the test of economic chaos is more difficult than its observation because the economic time series are characterized by strong noise, growing trend and time evolution. An important breakthrough is the usage of time-frequency representation by Chen [30, 31] in the analysis of stock markets. The author proposed the so-called "color-chaos model" to prove the existence of persistent chaotic cycles in the U.S. stock market. The characteristic frequencies of deterministic cycles were found and the relationship between frequency patterns and dynamical changes in business cycles provided an explanation for the stock market crash in October 1987. After Chen's work, many methods of analyzing chaotic time series were applied to stock price time series and the FMH was developed rapidly.

1.2 Stock Market Analysis

As a result of the above mentioned studies, not only has the limited predictability of the stock market been further substantiated, but the methods of modeling the stock market have also made great progress. There are two main approaches in the area of stock analysis, namely, the fundamental analysis and technical analysis. The fundamental analysis is a technique to evaluate a company and then make investment decisions by analyzing the fundamental factors that affect a company's value and future prospect. The assumption underlying is that stock prices do not really reflect the company's value in the short run, but it will eventually return to its real value. It is a powerful method for selecting individual stocks, understanding relevant industry group, and doing long-term investment. The other approach is the technical analysis. Unlike the fundamental analysis that studies the determinants of market movements, the technical analysis emphasizes on the behavior of the market itself which can be universally applicable to different stocks. It assumes that stock prices reflect all the information in the market and the price patterns will repeat themselves from time to time. The technical analysis is usually carried out in the form of charts, technical indicators, and oscillators. Due to its sensitivity to the market movement, it is widely used in short-term trading. In general, the fundamental analysis explains the reasons of stock price movements, while the technical analysis focuses on the time of entry and exit point. However, the fundamental and technical analyzes by themselves have drawbacks. For the fundamental analysis, the common criticisms are mainly focused on its inherent subjectivity, time-consuming analysis, too specific a model for some particular companies or industries, and too many economic variables involved. Drawbacks of the technical analysis include lack of theoretical basis, too narrow area of factor selection and unsuitability for long-term holding. Therefore, it is believed that the combination of these two techniques would be effective and efficient.

With the development of the modeling theory and methodology, much more information is concerned in the models for the analysis and prediction. There is a tendency that the boundary between technical and fundamental analysis becomes indistinguishable. In this way, we will not categorize a model as the fundamental or technical analysis but consider which fundamental data or technical data that are to be included. Fundamental data usually comprise elements from macroeconomy (such as interest rate, currencies, CPI, and PPI), industry sector, and the company itself (such as dividend payout, earning, growth, and profit margin). Technical data contain much less variables, only including the opening, closing, highest, lowest stock prices and volume. Through transforming and combining these two types of data, the derived data include the returns, volatility, turning points, artificial data, etc. Many models and approaches are available now in the literature for analyzing financial markets using these data. In the following, we give a general overview of models from various branches of science, which have been borrowed and utilized in the stock market analysis.

The booming of modeling methods has revealed many significant features and behavior of the market, which further stimulates the development of more advanced modeling techniques. Traditional economic models such as the capital asset pricing model (CAPM) [116] and the Black–Scholes model [19] serve as useful tools in pricing stocks, but they are found to be not suitable in analyzing complicated phenomena. As a result, theories and methods from other disciplines are gradually integrated with these economic models to enhance their performance. For example, models from time series analysis, physics, computational intelligence and systems theory are all powerful tools to facilitate the market analysis.

The traditional models in the time series analysis can roughly be categorized into two basic types: the univariate and multivariate models. In the univariate models, the autoregressive (AR) model, the moving average (MA) model and the combination of them, i.e., the autoregressive moving average (ARMA) model, are most commonly used. All these models are under the assumption that the time series are stationary stochastic processes. If ARMA is used to model the time series, whose difference is stationary, it is called the autoregressive integrated moving average (ARIMA) model [22]. These traditional univariate models assume that the time series reflects all the useful information including the influences of underlying explanatory variables. In order to investigate how the stock market correlates to other economic components, they are naturally expanded to be multivariate. One of the popularly used multivariate techniques is the multivariate autoregressive moving average (MARMA) model, which has advantages in forecasting marketing time series with explanatory variables [61]. The vector autoregressive (VAR) framework is another popular multivariate extension in capturing the dynamics between multiple time series. It is especially useful in measuring market responses to exogenous shocks [25]. Friedman and Shachmurove [49] used the VAR model to investigate the interdependence between stock markets of eight European Community countries, in which financial integration between the larger markets was found to be higher than that between the smaller ones. We note that all of the previously mentioned models assume the variance of the time series to be identically and independently distributed. However, time-varying variance, which is also called heteroscedasticity , exists in many financial time series including stock returns series. The autoregressive conditional heteroscedasticity (ARCH) and the generalized autoregressive conditional heteroscedasticity (GARCH) models [44] as well as their model family suggest a process to forecast this time-varying variance. In this sense, they have become widespread tools for dealing with the market risk management [67]. Nevertheless, because of their structures and imposed assumptions, these models are still inadequate for simulating the behavior of the whole market.

Benefiting from the development of multidisciplinary fields, theories and methodologies from physics, engineering, and even social science are integrated with economics in building new models for the stock market or financial markets in general in recent decades. The FMH is credited to the development in physics and engineering. In scientific sectors, analyzing time series in the frequency domain is widely used and accepted to find features that are unobservable in the time domain. The time-frequency analysis, which originates in quantum mechanics and acoustic physics [128], can be used to present the information of evolutionary time series in both the time and frequency domains simultaneously. It provides a new perspective and a powerful technique to fully describe the movement of stock prices or returns over time. The revelation of economic chaos and the resulting FMH have benefited a lot from this technology. It has once again testified the existence of trend and cycle components in economic and financial time series.

Another well-studied cluster is the computational intelligence in finance, which is represented by artificial neural networks (ANN) and support vector machines (SVM). With their capability of simulating complex nonlinear relationships between the input influential factors and the output, they both perform well in the prediction of stock prices. Additionally, ANNs largely facilitate the development of stock trading systems. Kimoto *et al.* [77] first constructed a stock trading system based on a modular neural network with supplementary learning algorithm. The results from Jang *et al.* [72] as well as Motiwalla and Wahab [99] were also quite satisfying. However, some inherent limitations of the ANN-based models, such as the need of a large amount of data in the training procedure, the overfitting and local minimum problems, restrict the application of ANNs in modeling the stock market. Established on the structural risk minimization principle to estimate a function, the SVM has been shown to be resistant to these inherent limitations that ANNs have. Thus, the SVM always achieves a better generalization performance than ANNs. One of the well-known studies regarding the usage of the SVM in the stock market prediction was presented by Kim [75]. He used the SVM to predict the direction of daily price changes in the Korea stock price index and compared the results with BP neural network and Case Based Reasoning (CBR). Yang [132] forecasted the Hang Seng Index and Dow Jones Industrial Average Index by using a support vector regression with non-fixed and asymmetrical margin setting and momentum. His model generates better results than using the radial basis function (RBF) neural networks, the GARCH model and the AR model. Although the SVMs can provide the global minimum as the solution, the results obtained through these non-parametric methods still lack transparency, since they tend to solve merely a convex optimization problem.

A breakthrough was contributed by system economics, a group of methods that analyze the financial market as a complex system. In 1980, Michael [97] pointed out some potential areas where economics may interact with systems theory, one of which is behavior finance, a very popular area of study today. It aims to examine psychological biases of investors and the consequent influences on the market. These ideas lead to the development of models that combine knowledge from economics, psychology, neuroscience and systems science. The agent-based model is one such useful tool for understanding the market microstructure. Poggio *et al.* [106] proposed a four-component repeated double-auction market model and conducted six experiments to investigate some key market dynamics and properties, which include the market efficiency, price deviation and the distribution of wealth. Such properties have also been investigated by LeBaron *et al.* [84, 85]. Chen and Yeh [33] focused on the belief and behavior of traders, while Chen and Liao [32] showed that the stock price-volume causal relation exists without any explicit assumptions. All of these results are based on an agent-based artificial stock market. As Orrell and McSharry [102] presented in their survey paper, system dynamics is another powerful and promising approach in analyzing the financial market. Considered as a tool for learning a complex world, system dynamics has found successful applications in a wide range of areas including the financial market analysis [35]. Cao and Wang [26] showed in their work how information technology, control, and computer technology can contribute to financial engineering. It should also be noted that Gerencsér [51] proposed a behavioral finance model based on the systems theory. In their model, the behavior of agents in the stock market was depicted by a closed-loop system where the plant was the market and the controller was the belief and behavior of agents. Although their model provides a new perspective for understanding the behavior of the stock market, it mainly focuses on the online regression of an autoregression (AR) model rather than the structure or the dynamics of the system.

1.3 Contribution of This Work

As mentioned earlier, there are vast of methodologies from various areas such as the traditional approaches in time series analysis, computational intelligence, and physics and engineering, which have been adopted to model the stock market. Most of these models focus on certain specific aspects of the market, but are inadequate for a comprehensive analysis of the market behavior due to their basic theoretical foundation and limitation. The essential drawbacks of these methods still prompt people to search for new ideas, especially techniques that can perform well in complicated situations. System economics is a popular and promising direction as it has provided many powerful tools in analyzing the market as a complex system, under which the features and structures of the market can be better explored. However, it would be more desirable to have a general framework, which can combine various foundations together and thus can provide more meaningful insights of the stock market. These problems pose a strong motivation for further research on modeling of the stock market using systems theory.

The research efforts in our monograph contribute to the existing literature both in theory and practice. The theoretical aspect of our contributions is the development of a comprehensive framework for modeling the stock market or financial markets in general from a viewpoint of system dynamics. More specifically, we have proposed a feedback adaptation structure to systematically model the stock market so that its market dynamics and properties can be better understood and captured. Under this framework, the modeling process is considered as identifying a dynamical system, in which the real market is treated as an unknown plant or system and the proposed identification model is tuned by feeding back the resulting matching errors. Like most of physical systems, a financial market also has its fast and slow dynamics which correspond to its external and internal forces. Our identification model consists of an internal model and an adaptive filter, respectively taking the slow and fast dynamics of the market prices into consideration. Our results show that the proposed framework gives the best prediction results as compared to the traditional methods such as the well-known ARMA model with exogenous input (ARMAX).

The working scheme of the proposed structure of modeling involves a crucial component, which is to identify the input influential factors. A selection method is proposed and is shown to have an ability of identifying the most appropriate influential factors. It provides an essential source to measure the market movement and reveals that the influential factors are frequency dependent and market dependent. We have investigated both developed and emerging markets including the U.S., China, Hong Kong, and Singapore markets. All the obtained results have demonstrated that our framework is effective and efficient in capturing significant market influential factors and predicting the market movement. We would like to emphasize that our proposed system adaptation framework does not depend on any particular model. Besides the models adopted in this monograph, others can also be utilized as the internal model and the adaptive filter as long as they are able of capturing the internal and external forces. Our framework integrates both fundamental and technical analyses by taking the advantage of the complementary nature of both approaches, under which market behavior like information feedback and dynamics of the system can be better identified and examined.

We have also obtained some very promising results by applying the proposed framework to determine major market turning periods. When analyzing the internal residue (the estimation error of the internal model) in the frequency domain, we discover that its frequency response exhibits strong characteristic patterns from time to time. As evidenced by statistical tests, the appearance of these characteristic patterns in the frequency response provides information on major turnings in the trend of stock price movements. We have proposed a set of rules to identify this kind of frequency patterns and then determine whether the market is in its major reversal. The forecasting rules have been successfully tested in the Dow Jones Industrial Average (DJIA) of New York Stock Exchange, the Composite Index of Shanghai Stock Exchange (SSE), the Hang Seng Index (HSI) of Hong Kong Stock Exchange and the Straits Times Index (STI) of Singapore Stock Exchange (SGX). Inspired by the stability property of a dynamical system in systems theory, we conduct a quantitative study of the relationship between the instability of the internal model obtained and the major turning periods in the market trend, which turns out to have greatly improved the accuracy in forecasting major market turnings. Although research along this line is still at a very preliminary stage, it has been proven, however, to provide an interesting and promising direction of applying systems theory to the analysis of financial markets.

To facilitate the analysis of the stock market, a MATLAB toolkit with a user-friendly graphical interface has been developed. This flexible and powerful toolkit not only integrates many popular methods in the technical analysis, but also includes the framework we proposed. It provides daily as well as real-time stock price data and is capable of performing various technical analyses, causality tests, market turning period forecasting and many others. All indicators provided in the toolkit have fully customizable parameters, which allow for greater flexibility. With the toolkit, users can easily carry out the examination of various trading rules without the need of in-depth programming or chart reading skills, even though a basic understanding of the technical analysis is required.

The utilization of systems theory in modeling the financial market is relatively new. As pointed out by Orrell and McSharry [102], a framework that can tie many of the foundations of systems economics together remains to be established. Our work contributes to this area by presenting a unique system adaptation framework to model the financial market as a complex system. Although we focus our attention on the stock market, the structure of the framework and the methods to establish the components involved can also be utilized to analyze other financial markets. In contrast to single dynamic models, our framework provides not only a more accurate prediction but a better description of the real market as well, under which the information flow of the real market can be clearly reflected in the hierarchy of the identified system. Since it is a new field of research, there are more works to be done to make the theory completed. One possible direction is to follow the system decomposition techniques of Chen *et. al.* [27] to further decompose the identified models under our framework into various subsystems, which, along with the interconnections that exist among them, would reveal more structural properties of the market. The system decomposition methodologies reported in Chen *et. al.* [27] have been proven to be a powerful tool in analyzing physical systems and instrumental in designing control systems for engineering problems. It is our hope that this monograph would serve as an entrance to this promising area.

1.4 Preview of Each Chapter

The remaining content of this book is divided into seven chapters. In Chapter 2, we present the structure and construction of our system adaptation framework. Viewing the stock market as a highly complex system, we propose a feedback adaptation structure based on systems and control theory to model the behavior of financial markets, or more specifically, the stock market from a dynamic system point of view. The proposed framework consists of an internal model and an adaptive filter, aiming to capture the slow and fast dynamics of the market, respectively. Feedback and force are two essential elements in the framework. The output–error model is adopted as the internal model whereas the adaptive filter is a time-varying state space model. The input-output behavior, and internal as well as external forces are then identified. The structural properties, model selection and related information flow of the framework are examined. Throughout this chapter to Chapter 4, the index of the Dow Jones Industrial Average (DJIA) is used as an example to illustrate the working scheme and the features of our framework. The global financial crisis in 2007 is particularly highlighted.

Chapter 3 focuses on the input selection of the system adaptation framework to identify crucial market influential factors. A selection method is proposed based on an empirical procedure and statistical tests. More specifically, we first carry out an empirical research to preselect influential factors from economic and sentiment aspects. The causal relationship between each of them and the internal residue of the market is then tested. Lastly, a multicollinearity test is applied to those factors that show significant causality to the internal residue of the market. The redundant factors will be excluded from our input selection.

Chapter 4 examines the forecasting capability of the system adaptation framework together with the appropriate input selected. The DJIA provides a successful case for demonstrating the great ability of our approach in understanding the dynamics of the stock market. We specifically investigate the DJIA from January 2008 to November 2011, the period right after the global financial crisis in 2007. The whole period is separated into four subperiods according to the economic situation. Our result shows that the system adaptation framework outperforms the existing methods, such as the commonly used ARMAX approach, in predicting stock prices. It is evident that with the carefully selected inputs, our framework is capable of providing excellent results in modeling the market dynamics.

To further test the new framework, we apply it to three Asian markets, namely, the stock markets of China, Hong Kong and Singapore. In Chapter 5, the complete process from the input selection, internal model and adaptive filter estimation, to market prediction is presented for these three markets. All the results demonstrate the effectiveness of our framework. Similar to the prediction of the DJIA, the new approach shows its great predictive ability especially in the complicated economic situations. We further investigate in this chapter the features of influential factors in general where particular attention is devoted to frequency and distinct influences in the markets.

An application of the system adaptation framework is given in Chapter 7. It focuses on forecasting the major turning periods in the market trend. In order to reveal some important properties that could not be observed in the time domain, we analyze the internal residue in the frequency domain. Tests find that a market trend is about to reverse when the internal residue begins to display certain characteristic frequency patterns in its power spectrum. We develop a set of rules to identify this kind of frequency patterns and determine if the market is experiencing a major turning. These rules work well for stock indices from U.S., China, Hong Kong and Singapore, in which most of the time our forecasting results of major market turnings are accurate. Structural changes in macroeconomic situations are also investigated, providing some possible explanations for our method. The result is further confirmed by some very interesting phenomenon. We have observed that there is a link between the market turning periods and the instability of the internal model under our framework, which in turn can be utilized to enhance the forecasting accuracy.

In Chapter 7, we document the design and functionalities of a MATLAB toolkit, namely, the toolkit for technical analysis of stocks (T-TAS). It is developed to perform a comprehensive analysis on the stock market data. Benefiting from the MATLAB GUI tools, the toolkit has an attractive and intuitive graphic user interface with advanced functionalities. It allows users to easily download real-time and historical stock data online from Yahoo Finance [131], to identify trading opportunities by many trading rules provided, and to analyze the characteristics of the stock. No profound knowledge on technical analysis or programming is required. The T-TAS is programmed to have greater flexibility for research purposes. The parameters of the technical indicators included in the toolkit are fully customizable.

Finally, in Chapter 8, we highlight some possible topics worth further research, which include the enhancement of the structure of the proposed system adaptation framework to capture more complete multiple time-scale market dynamics and to incorporate more market information, such as influential factors with behavior characteristics, and the further investigation of the interconnection between the system structural properties of the identified model and the market inherent features.

Chapter 2
A System Adaptation Framework

2.1 Introduction

A system refers to a configuration of components connected together by intertwined relationships which acts as a whole [12]. In the broadest sense of the thinking about systems, "everything is a system" [24]. To illustrate, in engineering, it is always referring to a physical system like a hardware or software framework; in science, a system could be a portion of the universe, a method or an algorithm; in social science, the human organizations, human activities, and artificial societies are also viewed as systems. If a system consists of a large number of interacting elements and appears a nonlinear and dynamic behavior, it is considered as a complex system. A special category of complex systems is known as complex adaptive systems that are able to adapt themselves to the changes of the surrounding environments. Examples can be found almost everywhere, including social systems, ecosystems, immune systems and financial systems. As an important part of the financial systems, the financial market is naturally considered as a highly complex adaptive system.

The study of complex systems involves multidisciplinary concepts, knowledge and methods. Parts, wholes and relationships are the basic questions it considers. For example, i) the interconnections and interactions between different components, ii) the relations and differences between the "integrated whole" and its components, and iii) the interactions between the whole system and its environment are the three key features need to be investigated. The theory of complex systems and its applications have attracted more and more attention nowadays. Inspired by the so-called system thinking, which is a style of systematic problem solving process, we propose in this chapter a system adaptation framework to model the stock market or financial market in general.

The first question one would naturally ask is how to simplify a real system. In fact, any mathematical model is a simplification of a system for certain specific purposes. A complex system involves a huge number of factors and variables that likely surpass the capacity of any already known heuristic system or calculating device, not to mention some parts of which can be unknown or cannot be modeled at all. In most of studies and applications, it is generally not necessary to identify

X. Zheng & B.M. Chen: *Stock Market Modeling and Forecasting*, LNCIS 442, pp. 13–27.
DOI: 10.1007/978-1-4471-5155-5_2

a model that characterizes all the information of the system. For our studies of the stock market or financial markets in general, a simplification of the market mechanism and market input–output relationship is considered when we construct our system adaptation framework.

In the modeling of a complex system, another challenging issue is how to quantitatively describe the system. In systems theory, people prefer to represent signals and systems in diagrams [2]. The block diagram representation, which is heavily used in the engineering context, is a useful tool for visualizing a system and analyzing its information loop. We will also use a block diagram to depict our proposed system adaptation framework and the design of its components.

We should note that Z-transform is frequently used in our modeling process. For a discrete-time signal $x(n)$, its Z-transform $X(z)$ is defined as

$$X(z) = \mathscr{Z}(x(n)) = \sum_{n=-\infty}^{\infty} x(n)z^{-n}, \tag{2.1}$$

where n is an integral and z is a complex variable.

In what follows, we present the structure of the system adaptation framework. The index of the Dow Jones Industrial Average (DJIA) is used as an example to illustrate the working scheme and features of the framework.

2.2 Structure of System Adaptation Framework

The essential idea and approaches of complex systems have offered novel insights on the modeling of financial markets. Viewing the stock market as a highly complex system, we propose a closed-loop adaptation framework based on the well-established systems theory. It provides a systematic way to characterize the market behavior. Inspired by ideas used in identifying physical systems, the stock market modeling process is treated as the problem of identifying a dynamic plant as shown in Figure 2.1. The input–output behavior of the stock market is represented by the identification model $\hat{S}$ with its output $\hat{p}$ being the estimated stock price. The actual stock price p is the output of the real stock market S. We note that both S and $\hat{S}$ have the same input r, which consists of external influential factors of the stock market. We aim to determine the model structure and parameters of $\hat{S}$ such that the error e, the differences between the actual and estimated stock prices, is minimized.

It is believed that information inside and outside the stock market would act as forces to regulate the share price. Thus, the signals processed in our framework are considered as market forces. In a nonlinear deterministic system, the irregularity of the system's behavior is attributed to both internal characters and external inputs. Similar characteristics are obtained from the theory of nonlinear dynamic economics, suggesting that the economic fluctuations are caused by both internal and external forces. In the stock market, the internal force is believed to be generated by some fundamental factors like market mechanism, company value, profitability of a company, etc., whereas the external force is usually constituted by external

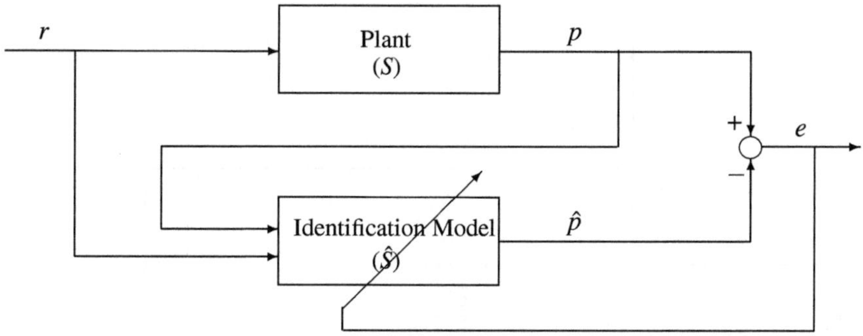

Fig. 2.1 The system adaptation framework

shocks from macroeconomic indicators, policies, global economic conditions, psychological biases, to name a few. These two kinds of forces act upon the stock market, making the market have slow and fast dynamics, respectively. As such, we construct a market model (as depicted in the block diagram of Figure 2.2) with an internal model I and an adaptive filter A to capture the slow and fast dynamics of the market, respectively. We should note that the symbol z in Figure 2.2 represents the usual Z-transform operator, i.e., it shifts an appropriate variable one step forward, whereas z^{-1} shifts a variable one step backward in time.

A time series always consists of trend and cycle components. Economic and financial time series have also been proven to have these characteristics by the time-frequency analysis. As such, the internal model, I, which can be regarded as a trend generator, is to estimate the market trend. It is rational to assume that the dynamics of the internal model change at a relatively slow pace, so that it can be approximated by some time-invariant systems. Note that the internal model is independent of the adaptive filter. It uses only actual historical prices to estimate the internal stock price $\hat{p}_{\mathrm{i}}$. For easy references, we define

$$e_{\mathrm{i}}(n) = p(n) - \hat{p}_{\mathrm{i}}(n), \tag{2.2}$$

as the *internal residue* in our framework. It is reasonable to assume that the internal residue is mainly due to external influences.

In the literature, information outside the stock market always accounts for external influences to the market. The adaptive filter A is thus introduced to capture such influences. It generates the estimated error $\hat{e}_{\mathrm{i}}$ of the next step by analyzing major influential factors of the stock market together with the historical information of the internal residue. Working as a cycle generator, this adaptive filter compensates the identification error e by capturing the fast dynamics of the market. It is assumed that the impact of each influential factor on the stock price movement is time dependent, so that their significances should vary at different stages. As such, it is natural for us to choose a time-varying adaptive filter linked to the external inputs. The application of models with time-varying coefficients in the analysis of economic

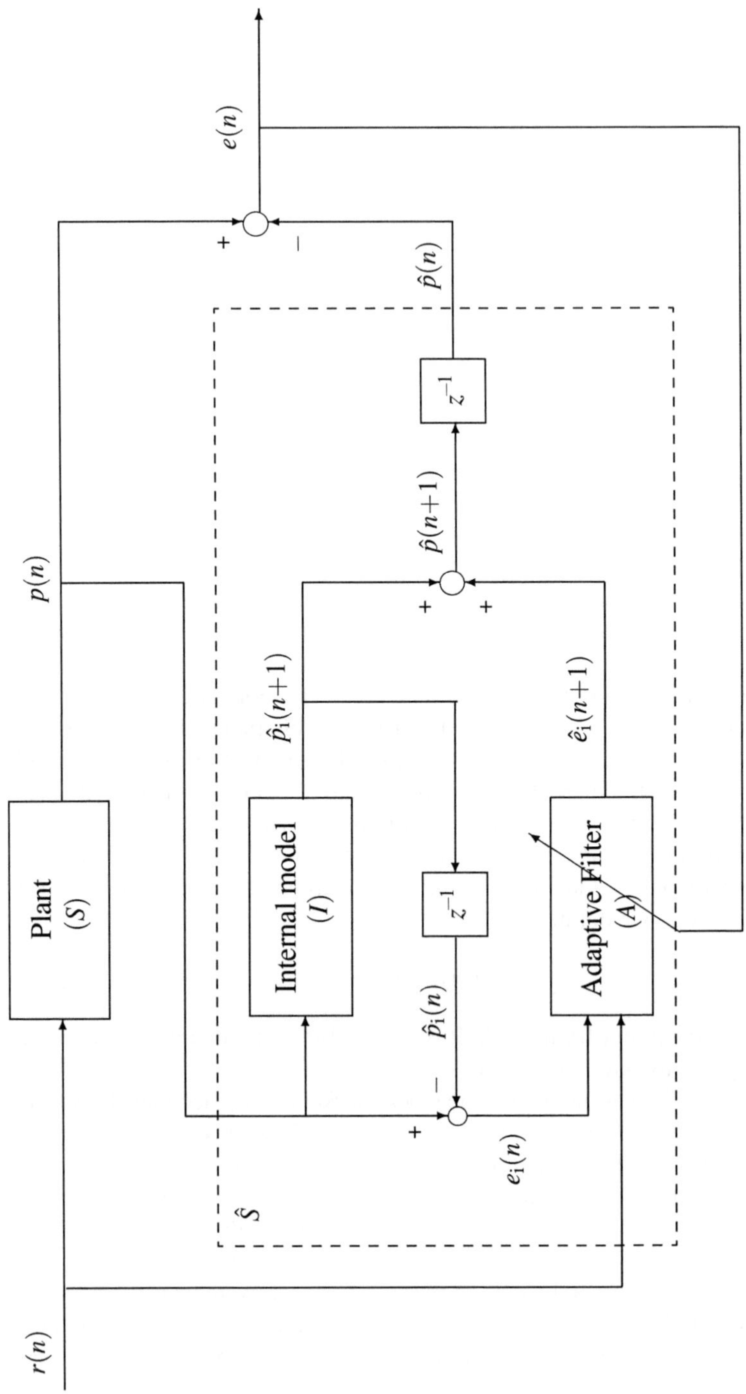

Fig. 2.2 Block diagram of the structure of system adaptation framework

and financial time series can be frequently found in the literature. Young [136] described many approaches to model economic time series based on the estimation of time-varying parameters. Binder and Merges [16] found that using time-varying coefficients based on the cluster regression can make a very high proportion of market volatility explainable.

Finally, the estimated stock price under the system adaptation framework is given by

$$\hat{p}(n+1) = \hat{p}_{\mathrm{i}}(n+1) + \hat{e}_{\mathrm{i}}(n+1), \tag{2.3}$$

which is a one-step-ahead prediction of the price movement in the stock market.

Another key feature in this framework is the feedback mechanism, a ubiquitous factor in both engineered and nature world. The idea of feedback has been heavily used in designing control systems for physical systems. It exists in the stock market both in terms of information and psychology, helping to select information and adjust trading strategies. That is how feedback works in a system as an adaptive response to changes. In the identification process, e is fed back to adjust the adaptive scheme of A. At the same time, feedback also exists inside the internal model to represent the effects of the historical data.

2.3 Internal Model Design

As depicted in Figure 2.3, the internal model consists of three parts. First, the historical stock price is preprocessed by the exponential moving average (EMA) model:

$$p_{\mathrm{ema}}(n) = \frac{2}{N+1}p(n) + \left(1 - \frac{2}{N+1}\right)p_{\mathrm{ema}}(n-1), \tag{2.4}$$

where N is an adjustable parameter denoting specific number of periods in the EMA, p and p_{ema} denote the actual and EMA prices, which are respectively the input and output of the EMA model. Through (2.4), the transfer function from $p(n)$ to $p_{\mathrm{ema}}(n)$, denoted by $H_{\mathrm{ema}}(z)$, can be expressed as

$$H_{\mathrm{ema}}(z) = \frac{P_{\mathrm{ema}}(z)}{P(z)} = \frac{\frac{2}{N+1}}{1 - \left(1 - \frac{2}{N+1}\right)z^{-1}}. \tag{2.5}$$

Based on the data obtained from the EMA preprocess, an output-error (OE) model of multi-inputs and single-output (MISO) is employed to model the inherent evolution of stock prices. Its input $u_{\mathrm{oe}}(n)$ consists of current and $k-1$ previous samples of the EMA prices, i.e.,

$$u_{\mathrm{oe}}(n) = \begin{bmatrix} u_{\mathrm{oe},1}(n) \\ u_{\mathrm{oe},2}(n) \\ \vdots \\ u_{\mathrm{oe},k}(n) \end{bmatrix} = \begin{bmatrix} p_{\mathrm{ema}}(n) \\ p_{\mathrm{ema}}(n-1) \\ \vdots \\ p_{\mathrm{ema}}(n-k+1) \end{bmatrix}. \tag{2.6}$$

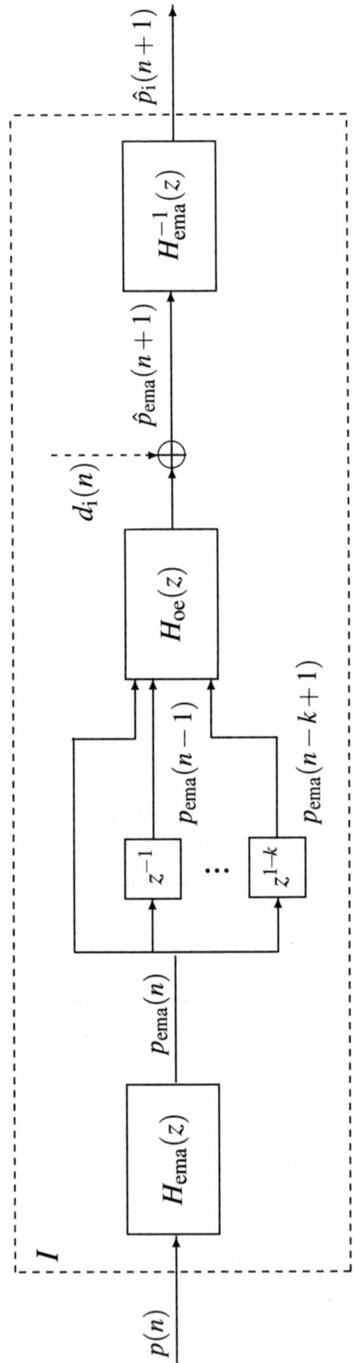

Fig. 2.3 The internal model

The transfer function of this MISO OE model, $H_{oe}(z)$, is characterized by

$$H_{oe}(z) = \left[H_{oe,1}(z) \; H_{oe,2}(z) \; \cdots \; H_{oe,k}(z) \right], \tag{2.7}$$

where for $j = 1, 2, \cdots, k$, $H_{oe,j}(z)$ is the transfer function on the j-th channel of the OE model, a rational function of z, and is given as

$$H_{oe,j}(z) = \frac{\Phi_j(z)}{\Psi_j(z)} \tag{2.8}$$

with

$$\Phi_j(z) = \phi_{j,1} + \phi_{j,2} z^{-1} + \cdots + \phi_{j,n_\Phi} z^{-n_\Phi + 1} \tag{2.9}$$

and

$$\Psi_j(z) = 1 + \psi_{j,1} z^{-1} + \cdots + \psi_{j,n_\Psi} z^{-n_\Psi}. \tag{2.10}$$

We also consider the system has a disturbance $d_i(n)$, which is assumed to be white noises, then the output processed after the OE model, i.e., the estimation of the EMA price, is given by

$$z\hat{P}_{ema}(z) = H_{oe}(z) U_{oe}(z) + D_i(z), \tag{2.11}$$

where $z\hat{P}_{ema}(z)$ and $U_{oe}(z)$ are the Z-transforms of $\hat{p}_{ema}(n+1)$ and $u_{oe}(n)$, respectively. Finally, the internal price $\hat{p}_i(n+1)$ can be derived by transforming the estimated EMA price $\hat{p}_{ema}(n+1)$ back with $H_{ema}^{-1}(z)$.

We would like to note that the OE model given above is not necessarily the best choice for the internal model in our framework. In fact, any model that is capable of capturing the market trends, i.e., the slow dynamics of the stock market, can be incorporated into our structure to yield a better result. The research for a more efficient and effective internal model is still an open problem.

In the following, we illustrate the procedure for parameter identification of the adopted OE model using the prediction error method reported in Ljung [88]. Let the estimation error of the OE model be denoted by

$$e_{oe}(n) = p_{ema}(n) - \hat{p}_{ema}(n). \tag{2.12}$$

Through (2.6) to (2.11), we have

$$\begin{aligned} e_{oe}(n) &= p_{ema}(n) - H_{oe}(z) u_{oe}(n-1) \\ &= p_{ema}(n) - \sum_{j=1}^{k} H_{oe,j}(z) p_{ema}(n-j). \end{aligned} \tag{2.13}$$

Define parameter vector θ as

$$\theta = \left[\theta_1 \; \theta_2 \; \cdots \; \theta_k \right], \tag{2.14}$$

where

$$\theta_j = \left[\psi_{j,1} \ \psi_{j,2} \ \cdots \ \psi_{j,n_\Psi} \ \phi_{j,1} \ \phi_{j,2} \ \cdots \ \phi_{j,n_\Phi} \right]. \quad (2.15)$$

Then, $e_{\mathrm{oe}}(n)$ in (2.13) can be written as

$$e_{\mathrm{oe}}(n) = p_{\mathrm{ema}}(n) - f\left(\theta, u_{\mathrm{oe}}(n-1)\right), \quad (2.16)$$

where $f\left(\theta, u_{\mathrm{oe}}(n-1)\right)$ is the function of θ and $u_{\mathrm{oe}}(n-1)$. The parameter vector θ can be estimated by minimizing the following cost function

$$V_K(\theta) = \sum_{n=1}^{K} e_{\mathrm{oe}}^2(n) = \sum_{n=1}^{K} \left[p_{\mathrm{ema}}(n) - f\left(\theta, u_{\mathrm{oe}}(n-1)\right)\right]^2, \quad (2.17)$$

where K is the sample size. Obviously, $V_K(\theta)$ is nonlinear with respect to θ. A possible solution is to use Newton–Raphson method to iteratively minimize this cost function $V_K(\theta)$. The estimation process is then given by

$$\hat{\theta}_K^{(i+1)} = \hat{\theta}_K^{(i)} + \mu_K \left[\frac{\partial^2 V_K\left(\hat{\theta}_K^{(i)}\right)}{\partial \theta^2} \right]^{-1} \frac{\partial V_K\left(\hat{\theta}_K^{(i)}\right)}{\partial \theta}, \quad (2.18)$$

where μ_K is the step size and $\partial^2 V_K\left(\hat{\theta}_K^{(i)}\right) / \partial \theta^2$ is the Hessian matrix that gives the search direction. The detailed algorithm of the prediction error method can be found [88].

2.4 Adaptive Filter Design

As depicted in Figure 2.4, in order to identify the dynamic influences of exogenous factors on the stock market, a time-varying state space model with instrumental variables is used as the adaptive filter. We consider the following general time-varying model [123, 137] to represent the input–output relationship of the adaptive filter:

$$Y(z) = \left[\frac{B_1(z,n)}{A(z,n)} \ \frac{B_2(z,n)}{A(z,n)} \ \cdots \ \frac{B_m(z,n)}{A(z,n)} \right] U(z) + \varepsilon(z), \quad (2.19)$$

where $Y(z)$ and $U(z)$ denote the output and input, respectively; $\varepsilon(z)$ denotes noise; and

$$B_j(z,n) = b_{j,0}(n) + b_{j,1}(n)z^{-1} + \cdots + b_{j,n_j}(n)z^{-n_j} \quad (2.20)$$

$$A(z,n) = 1 + a_1(n)z^{-1} + \cdots + a_{n_a}(n)z^{-n_a}. \quad (2.21)$$

As we have $\hat{e}_{\mathrm{i}}(n+1)$ being the output of the adaptive filter, and historical data of e_{i} as well as influential factors r with a lag length of δ being the input, the time-varying model we adopt is

$$\hat{e}_{\rm i}(n+1) = -\sum_{j=1}^{n_{\rm a}} a_j(n)e_{\rm i}(n-j+1) + \sum_{j=0}^{n_1} b_{1,j}(n)r_1(n-\delta_1-j) + \cdots$$
$$\cdots + \sum_{j=0}^{n_m} b_{m,j}(n)r_m(n-\delta_m-j) + \sum_{j=1}^{n_{\rm a}} a_j(n)\varepsilon(n-j) + \varepsilon(n) \qquad (2.22)$$

where $\hat{e}_{\rm i}(n+1)$ is the output of the adaptive filter; r_j denotes a selected influential factor with a lag length of δ_j; ε is a Gaussian white noise representing uncertainties; a_j, $j = 1,2,\cdots,n_{\rm a}$, $b_{i,j}$, $i = 1,2,\cdots,m$, $j = 0,1,\cdots,n_i$, are time-dependent parameters to be identified. It is fair to assume that all these time-varying parameters are statistically independent. We note that such a model is an extension of the autoregressive exogenous variable model by allowing the noise to be colored. The dynamics of such a force generation system are represented by those time-varying parameters, a_j, $j = 1,2,\cdots,n_{\rm a}$, $b_{i,j}$, $i = 1,2,\cdots,m$, $j = 0,1,\cdots,n_i$, which are assumed to evolve according to a random walk. For each of these time-varying parameters, it is assumed to be characterized by the following general stochastic model:

$$x(n) = x(n-1) + \eta_x(n), \qquad (2.23)$$

where x represents a time-varying parameter to be identified, and η_x is a white noise input associated with x, which is assumed to have a normal distribution $N(0, Q_x)$.

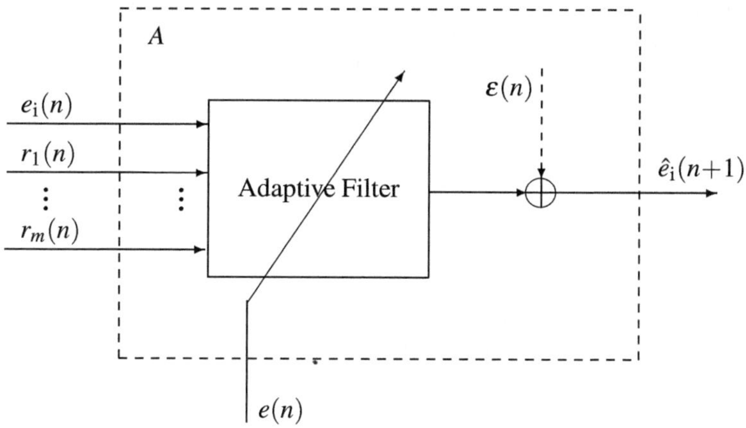

Fig. 2.4 The adaptive filter

By defining these parameters as state variables, we convert the model in (2.22) into a state-space form, which can be easily identified by some well-established approaches, e.g., the Kalman filters. Constructing vectors $X(n)$ and $H(n)$ as

$$X(n) = \left[a_1(n)\ a_2(n)\ \cdots\ a_{n_{\rm a}}(n)\ b_{1,0}(n)\ \cdots\ b_{1,n_1}(n)\ \cdots\ b_{m,0}(n)\ \cdots\ b_{m,n_m}(n) \right]^{\rm T}$$

and

$$H(n) = \Big[-e_\mathrm{i}(n)\ -e_\mathrm{i}(n-1)\ \cdots\ -e_\mathrm{i}(n-n_\mathrm{a}+1)\ r_1(n-\delta_1)\ \cdots\ r_1(n-\delta_1-n_1)$$
$$\cdots\ r_m(n-\delta_m)\ \cdots\ r_m(n-\delta_m-n_m) \Big], \tag{2.24}$$

we have the state space model:

$$X(n) = X(n-1) + \eta(n), \qquad \eta(n) \sim N(0, Q) \tag{2.25}$$

and

$$\hat{e}_\mathrm{i}(n+1) = H(n)X(n) + \mu(n), \qquad \mu(n) \sim N(0, \sigma^2). \tag{2.26}$$

where η consists of all the white noise inputs associated with the parameters to be identified, and

$$\mu(n) = \sum_{j=1}^{n_\mathrm{a}} a_j(n)\varepsilon(n-j) + \varepsilon(n). \tag{2.27}$$

Obviously, Q is a diagonal matrix with its diagonal elements being the variance of the related input noise. We aim to estimate $X(n)$, denoted by $\hat{X}(n)$, such that the resulting identification error $e(n) = e_\mathrm{i}(n) - \hat{e}_\mathrm{i}(n)$ is minimized.

Since the regressors include lagged terms of the output of the adaptive filter, which may correlate with $\varepsilon(n)$, we follow the idea of [137] to introduce an instrumental variable $u(n)$ for eliminating possible estimation bias:

$$u(n) = -\sum_{j=1}^{n_\mathrm{a}} \hat{a}_j(n-1)u(n-j) + \sum_{j=0}^{n_1} \hat{b}_{1,j}(n-1)r_1(n-\delta_1-j-1) + \cdots$$
$$\cdots + \sum_{j=0}^{n_m} \hat{b}_{m,j}(n-1)r_m(n-\delta_m-j-1). \tag{2.28}$$

Such an instrumental variable is highly correlated with the original regressor vector but uncorrelated with $\varepsilon(n)$. Under this setting, we construct the instrumental vector $\hat{H}(n)$ as

$$\hat{H}(n) = \Big[-u(n)\ -u(n-1)\ \cdots\ -u(n-n_\mathrm{a}+1)\ r_1(n-\delta_1)\ \cdots\ r_1(n-\delta_1-n_1)$$
$$\cdots\ r_m(n-\delta_m)\ \cdots\ r_m(n-\delta_m-n_m) \Big], \tag{2.29}$$

and we have

$$\hat{X}(n) = [\,\hat{a}_1(n)\ \hat{a}_2(n) \cdots \hat{a}_{n_\mathrm{a}}(n)\ \hat{b}_{1,0}(n) \cdots \hat{b}_{1,n_1}(n) \cdots \hat{b}_{m,0}(n) \cdots \hat{b}_{m,n_m}(n)\,]^\mathrm{T}. \tag{2.30}$$

From (2.25), it is obvious that the unknown parameters, or hyperparameters, in the covariance matrix Q determine the variations of the state variables. As such, we first carry out a hyperparameter optimization process, and then use the Kalman filter to perform the recursive estimation and prediction. We adopt the advanced maximum likelihood (ML) method [62, 137] to estimate the hyperparameters based on a set of historical data. The major innovation of this method is the introduction of the noise variance ratio (NVR) matrix Q_{r} and $\hat{P}$, which are respectively defined as

$$Q_{\mathrm{r}} = \frac{Q}{\sigma^2} \quad \text{and} \quad \hat{P} = \frac{P}{\sigma^2}, \tag{2.31}$$

where P is the prediction error covariance matrix associated with the estimated state vector $\hat{X}$. In what follows, we adopt the notation $(n|n-1)$ to denote the estimation of a specific variable in step n conditional on information up to step $n-1$. In this way,

$$\begin{aligned} \hat{P}(n|n-1) &= \frac{P(n|n-1)}{\sigma^2} \\ &= \frac{1}{\sigma^2} E\left[(\hat{X}(n) - \hat{X}(n|n-1))(\hat{X}(n) - \hat{X}(n|n-1))^{\mathrm{T}}\right]. \end{aligned} \tag{2.32}$$

Given a set of historical data $e_{\mathrm{i}}(1), e_{\mathrm{i}}(2), \cdots, e_{\mathrm{i}}(\tau)$, the Log-likelihood function of $e_{\mathrm{i}}(\tau+1), e_{\mathrm{i}}(\tau+2), \cdots, e_{\mathrm{i}}(K)$, conditional on previous information is calculated via commonly used prediction error decomposition by

$$\begin{aligned} &\log L\left(e_{\mathrm{i}}(\tau+1), \cdots, e_{\mathrm{i}}(K) \middle| e_{\mathrm{i}}(1), \cdots, e_{\mathrm{i}}(\tau)\right) \\ &= \frac{-(K-\tau)}{2} \log 2\pi - \frac{1}{2} \sum_{n=\tau+1}^{K} \log |\, var(e(n)) \,| - \frac{1}{2} \sum_{n=\tau+1}^{K} \frac{e^2(n)}{var(e(n))}, \end{aligned} \tag{2.33}$$

where $e(n) = e_{\mathrm{i}}(n) - H(n-1)\hat{X}(n-1|n-2)$ is also the one-step-ahead prediction error with the instrumental variables and $var(e(n))$ is its variance given by

$$var(e(n)) = \sigma^2 \left[1 + \hat{H}(n)\hat{P}(n|n-1)\hat{H}^{\mathrm{T}}(n)\right]. \tag{2.34}$$

The Log-likelihood function needed to be maximized is

$$\begin{aligned} \log L(\cdot) = \frac{-(K-\tau)}{2} \log 2\pi - \frac{K-\tau}{2} \log \sigma^2 - \frac{1}{2} \sum_{n=\tau+1}^{K} \log\left[1 + \hat{H}(n)\hat{P}(n|n-1)\hat{H}^{\mathrm{T}}(n)\right] \\ - \frac{1}{2\sigma^2} \sum_{n=\tau+1}^{K} \frac{e^2(n)}{1 + \hat{H}(n)\hat{P}(n|n-1)\hat{H}^{\mathrm{T}}(n)}. \end{aligned} \tag{2.35}$$

Partially differentiating (2.35) with respect to σ^2, conditional on given information, it can be shown that the estimation of σ^2, i.e., $\hat{\sigma}^2$, is given as

$$\hat{\sigma}^2 = \frac{1}{K-\tau} \sum_{n=\tau+1}^{K} \frac{e^2(n)}{1 + \hat{H}(n)\hat{P}(n|n-1)\hat{H}^{\mathrm{T}}(n)}. \tag{2.36}$$

Substituting (2.36) into (2.35) and removing the constant term as well as the negative sign, we obtain the following compact form

$$\log\hat{L}(\cdot) = \sum_{n=\tau+1}^{K} \log\left[1+\hat{H}(n)\hat{P}(n|n-1)\hat{H}^{\mathrm{T}}(n)\right] + (K-\tau)\log\left[\frac{1}{K-\tau}\sum_{n=\tau+1}^{K}\frac{e^2(n)}{1+\hat{H}(n)\hat{P}(n|n-1)\hat{H}^{\mathrm{T}}(n)}\right]. \quad (2.37)$$

The hyperparameters can be estimated by minimizing (2.37). The optimization algorithm we used is the Nelder-Mead simplex direct search method [81]. With the estimated hyperparameters, the Kalman filter can be use to provide recursive prediction and updating processes, in which the identification error $e(n)$ is fed back to tune the states. The following is the recursive algorithm:

Prediction:

$$\hat{X}(n|n-1) = \hat{X}(n-1|n-1) \quad (2.38)$$

and

$$\hat{P}(n|n-1) = \hat{P}(n-1|n-1) + Q_{\mathrm{r}} \quad (2.39)$$

Updating:

$$\hat{X}(n|n) = \hat{X}(n|n-1) + \hat{P}(n|n-1)\hat{H}^{\mathrm{T}}(n)\left[1+\hat{H}(n)\hat{P}(n|n-1)\hat{H}^{\mathrm{T}}(n)\right]^{-1} e(n) \quad (2.40)$$

and

$$\hat{P}(n|n) = \hat{P}(n|n-1) - \hat{P}(n|n-1)\hat{H}^{\mathrm{T}}(n)\left[1+\hat{H}(n)\hat{P}(n|n-1)\hat{H}^{\mathrm{T}}(n)\right]^{-1}\hat{H}(n)\hat{P}(n|n-1) \quad (2.41)$$

Finally, we note that the adaptive filter model presented above is by no mean the best choice. Similarly, as its internal model counterpart, the search for a more efficient and effective adaptive filter, i.e., a model that is capable of capturing the fast dynamics of the market, is also an open problem.

2.5 Case Study: Dow Jones Industrial Average

In this section, we use the Dow Jones Industrial Average (DJIA) as an example to illustrate the working scheme of our proposed system adaptation framework. More specifically, we present in the following data description of the DJIA for the time intervals of interest and the resulting internal model. As the adaptive filter is heavily associated with external influential factors, the corresponding result for the DJIA will be highlighted in detail in Chapter 3. As mentioned earlier, due to its importance as the snapshot of the health of the U.S. market and its widespread influence to the worldwide financial markets, the DJIA will be used for illustration for all the results presented in Chapters 2 to 4.

2.5.1 *Data Description*

Data of the stock market, especially those shortly before, during and after the 2007 U.S. sub-prime financial crisis, provide a valuable sample for both investors and researchers. Investigations of the market behavior during this period can help us gaining better understanding of the major influential factors of the market, the interactions between the market and these factors, and most importantly the market behavior if it is confronted with similar situations in the future. We use the daily closing prices of the DJIA from January 2008 to November 2011, as shown in Figure 2.5, the period from the beginning of the 2007 U.S. sub-prime financial crisis to the very recent past. We should note that all the stock market data are obtained from Yahoo Finance [131].

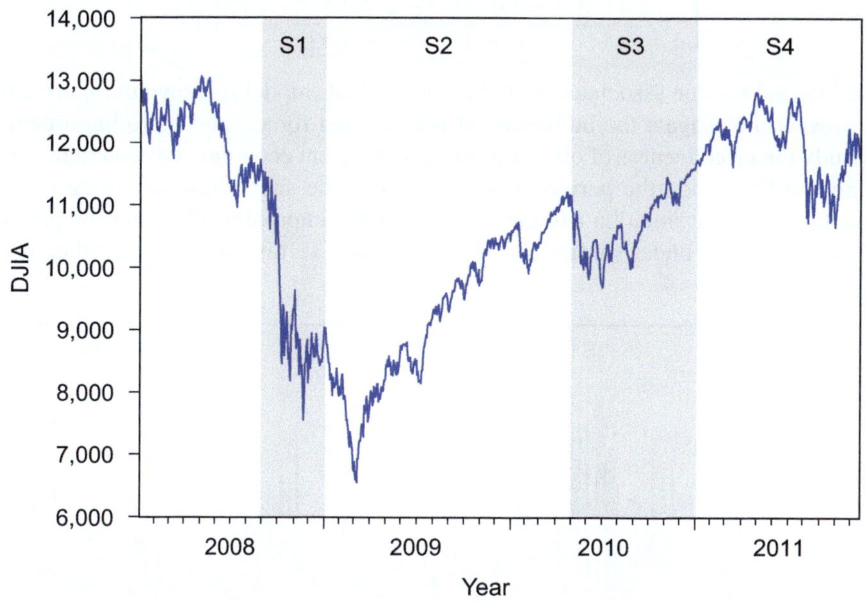

Fig. 2.5 Daily closing prices of the DJIA from January 2008 to November 2011

2.5.2 *Internal Model Estimation*

It was found in Chen [30, 31] that the persistent chaotic cycles in the U.S. stock market are around three to five years. We thus use five-year daily closing prices of the DJIA prior to the starting point of the testing period as a training set to tune the parameters of the internal OE model. For the period of interest in our studies, we make use of the DJIA daily closing prices from January 2003 to December 2007 for the purpose of identifying the internal model. In deriving the internal model of the DJIA, we choose an OE model with three input channels, i.e., $k = 3$, and select the adjustable parameter of the EMA preprocess $N = 12$, a popular choice used to create the Moving Average Convergence/Divergence (MACD) indicator in the financial market analysis. The estimation process starts with zero initial states and stops when the improvement is less than 10^{-4}. Particularly, this is a general rule that we adopt for estimating the OE model in our framework. It is also used in identifying market turning periods to be presented in Chapter 7.

The corresponding OE model for the DJIA during the period of interest is then obtained as

$$H_{\mathrm{DJIA}}(z) = \begin{bmatrix} \dfrac{0.9574z^{-1} - 0.5034z^{-2} - 0.321z^{-3} + 0.565z^{-4}}{1 - 1.329z^{-1} + 0.7312z^{-2}} \\ \dfrac{-5.127z^{-1} + 2.086z^{-2} + 0.3914z^{-3} + 0.8506z^{-4}}{1 - 0.62z^{-1} - 0.1239z^{-2}} \\ \dfrac{2.417z^{-1} + 3.049z^{-2} - 2.298z^{-3} - 1.82z^{-4}}{1 - 0.3215z^{-1} - 0.4643z^{-2}} \end{bmatrix}^{\mathrm{T}}. \tag{2.42}$$

The internal residue associated with the obtained OE model is shown in Figure 2.6. In order to investigate the influences of the external forces, and more importantly to study the effectiveness of our framework in different economic environments, we might further divide the period of interest into some subperiods according to the specific economic situation and the variance of the internal residue. For the period of the DJIA data under studies and for future use, we divide the market data into

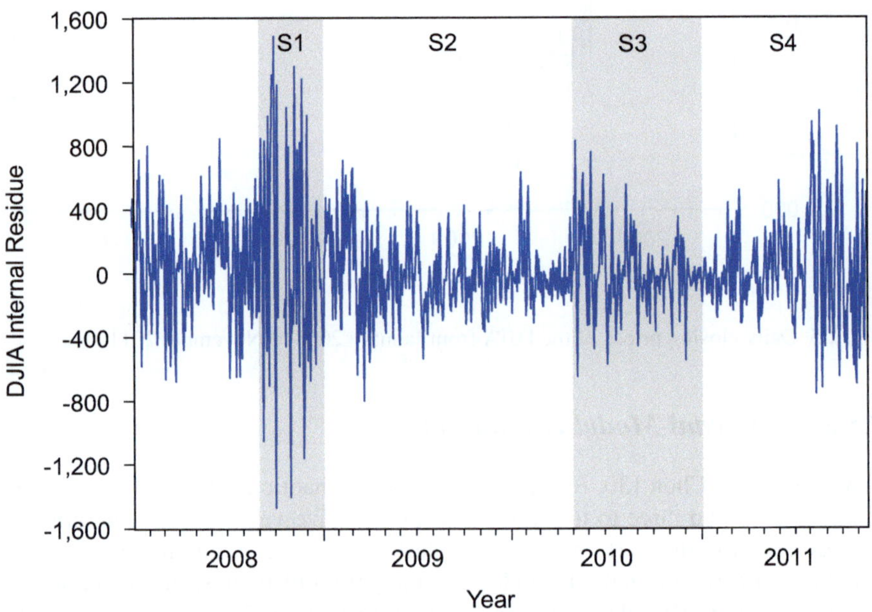

Fig. 2.6 Internal residue of the DJIA from January 2008 to November 2011

four subperiods (as shown in Figures 2.5 and 2.6) by observing the corresponding internal residue, and to highlight the performance of our proposed framework during the 2007 U.S. sub-prime financial crisis. As shown in Figure 2.6, Subperiod S1 is referred to the time interval from September to December 2008; Subperiod S2 from January 2009 to April 2010; Subperiod S3 from May to December 2010; and lastly, Subperiod S4 lasts from January to November 2011. It is easy to observe that in Subperiod S1, which starts from the month when the financial crisis hit its most critical stage, the variance of the corresponding internal residue is extremely large. After becoming relatively small in Subperiod S2, the variance of the internal residue increases again in May 2010, when is the beginning of Subperiod S3. It becomes much larger during Subperiod S4, which corresponds to the downturn of the market in 2011. These subperiods will be used to examine the related economic situation and the performance of our framework later in Chapter 4.

Chapter 3
Market Input Analysis

3.1 Introduction

Unlike most of physical systems, especially engineering systems, in which the system inputs are generally well-defined and structured to meet certain specifications and requirements, the stock market has too many input factors that might influence its internal behaviors and hence its outcomes. On the other hand, the output of the stock market, i.e., the market closing price used in our modeling process, is fixed once the market is closed and is relatively clean compared to its counterpart in the engineering systems, in which their outputs are normally coupled with measurement noises. This chapter focuses on the selection of appropriate input influential factors for the system adaptation framework under study.

In the stock market, it is common to have a mixture of information, some quantitative, some qualitative and some that cannot be measured at all through common sense. Additionally, since the market follows the law of demand and supply, market fluctuations are affected by the collective behavior of the investors. In such a system, the feedback effect may be more complicated than or even completely different from the usual physical systems. Hence, identifying the external information that affects the market is essential to modeling, as it represents the interaction of the whole system with its environment.

The input selection is one of the key elements in our framework, which in fact may become even more decisive in certain situations. The fast changing external force is generated by the information outside the stock market such as the economic, fundamental and various other influential factors. These influential factors work as the input to our framework to regulate stock prices through the adaptive filter. As their effect on the market changes from time to time, identifying the most influential factors in the given period of time is critical to yield good predicting performance. As mentioned earlier, the information contained in the historical stock prices form the internal force, which is reflected in the internal model, representing the slow dynamics of the price series. The external force is much more complicated compared to its internal counterpart, requiring a time-variant adaptive filter to capture its fast changing dynamics.

X. Zheng & B.M. Chen: *Stock Market Modeling and Forecasting*, LNCIS 442, pp. 29–42.
DOI: 10.1007/978-1-4471-5155-5_3

In this chapter, we propose some new forms of market influential factors together with a procedure to identify the most crucial input elements for the system adaptation framework. It involves an empirical selection process and then follows by statistical tests. Once again, the U.S. stock market, more specifically, the DJIA, is taken as an illustrative example to demonstrate the practicality of our selection process. The investigation of other markets including those from China, Hong Kong and Singapore will be reported later in Chapter 5.

3.2 Influential Factor Selection

Many results can be found in the literature concerning the influential factors of the stock market, most of which come from economic and sentiment aspects. Economic indicators are generally used to judge the well-being of the economy and predict its future performance. They fundamentally determine the movement of stock prices as the stock market has become an increasingly important component of the economy. Interest rate, inflation rate, money supply and commodity price are all widely accepted as economic indicators. A famous and one of the earliest studies of this topic is from Chen *et al.* [29], the results of which indicated that a set of economic variables including industrial production, interest rate, inflation and oil price are important in explaining expected stock returns. Bodurtha *et al.* [20], Campbell and Ammer [25], Binder and Merges [16], Kim [66] and Rapach *et al.* [108] have done similar studies. They investigated the predictive power of future dividends, discount rates, price-level uncertainty and ratio of expected profits to expected revenues.

Besides economic indicators, investor sentiment becomes a crucial factor in recent studies. Sentiment indicators act as a measurement of the situation of demand and supply, representing the general opinion of the investors towards the market. Their contributions to short-term variations are particularly significant. Baker and Wurgler [10, 11] provided a general method to measure investor sentiment; by their approach they have demonstrated how sentiment affects the stock market as a whole. In addition, the reactions of the stock market to noise trader risk [86], aggregate earning news [79], aviation disasters [41] and even terrorism activities [74] have also been investigated. All these works are based on some sentiment indicators, among which the Chicago Board Options Exchange Volatility Index (VIX) is a very common one used for measuring the fear of investors.

Based on some empirical research, we first select a set of key influential factors, which include both economic and sentiment indicators, and perform necessary preprocessing to reform the selected indicators such that the resulting data will be better fit to our proposed framework. After which, we then apply a series of statistical tests, which include linear time-varying and/or nonlinear causality tests as well as multicollinearity tests, on each selected indicator to detect its causality relationship with the internal residue. More specifically, a linear test is first conducted on each input–output pair of the adaptive filter, i.e., the influential factor and the internal residue. A nonlinear causality test is to be further carried out if no significant causality is displayed in the linear test. Finally, a multicollinearity test is then adopted to remove redundant indicators.

3.2.1 Causality Tests

There are a vast member of causality tests reported in the literature that identify the causal relationship between the stock price and the market influential factors, of which the original and also best-established approach is the Granger causality test [58]. Generally speaking, the Granger causality test is within the linear regression context. It provides useful information on whether the past knowledge of an influential factor could significantly improve the short-term prediction of stock prices, and vice versa. The so-called standard F-test was designed for stationary series. The technique was enhanced and popularized by the study of Sims [120], in which Sims proposed an ad hoc prefilter to stationarize the series in analysis, and then the causal relationship between money and income in the U.S. market was examined. The enhanced test has been widely applied in various areas, especially in empirical macroeconomics and empirical finance.

However, causality patterns may change from time to time due to various reasons such as fast changing dynamics in the economic environment. As such, the technique has been further extended for those dynamic models with time-varying parameters. Geweke [52, 53] quantified the causality in the form of linear dependence between signals based on VAR models. Such a concept can be easily extended to test dynamic models. For example, Thoma [124] used rolling windows Granger causality tests to document instabilities in the causal relationship between monetary policy indicators and future output. An alternative method in testing time-varying causality is to make use of Markov regime-switching models. By using an unobservable finite Markov chain to allow for changes in causal links over the sampling period, Psaradakis *et al.* [107] investigated the U.S. money-output relationship based on a VAR model with time-varying parameters. Lo and Piger [91] used a similar method to detect the changing causality linkage in the response of U.S. output to a monetary policy action. The time-varying causality test has also been frequently reported in the literature of neuroscience. Ding *et al.* [40] used a short-window spectral analysis to construct a time-variant Granger causality test, but it requires the stationarity of signal within the short-time window. Hesse *et al.* [63] loosened this requirement by recursively fitting a VAR model with time-dependent parameters, and then the time-variant Granger causality strength was calculated. Similar work could be found in Roebroeck *et al.* [110] as well as Bressler and Seth [23]. In our input selection process, we adopt the idea in the time-varying causality test.

Traditional Granger causality test only considers linear relationships which are nearly incapable of digging out the nonlinear dynamics in time series. As noted by Granger [59], Hsieh [68] and many others, the nonlinearity is an intrinsic and fundamental feature in financial time series. Baek and Brock [7] proposed a nonparametric statistical method to identify the nonlinear Granger causality. By allowing the series in testing to display short-term temporal dependence, Hiemstra and

Jones [65] modified the Baek and Brock test to discover the significant bidirectional nonlinear causality between the daily returns of the DJIA and the percentage changes in the New York Stock Exchange trading volume, for which the linear Granger causality test fails to discover the relationship. The work popularized the research in testing nonlinear Granger causality and the modified Baek and Brock test has become a commonly used method in economics and finance. Examples include causal relationships between international stock markets [13, 56], stock price and volume [119], stock price dividend relationships [73], currency markets, futures and cash markets [42]. We too adopt this modified Baek and Brock test in our input selection process. In what follows, we highlight some of the causality test techniques used in our selection procedure.

3.2.1.1 Time-Varying Causality Test

Hesse *et al.* [63] suggests a time-varying causality test which calculates the causality strength at each time point and then compare them with related threshold values. Considering the input and output time series r and e_{i}, which are, respectively, characterized by the following AR and bivariate AR models as

$$r(n) = \sum_{i=1}^{q_{\mathrm{r}}} \alpha_{1,i}(n) r(n-i) + \vartheta_1(n), \quad \Sigma_{11}(n) = var(\vartheta_1(n)), \tag{3.1}$$

$$e_{\mathrm{i}}(n) = \sum_{i=1}^{q_{\mathrm{e}}} \beta_{1,i}(n) e_{\mathrm{i}}(n-i) + \upsilon_1(n), \quad \Sigma_{21}(n) = var(\upsilon_1(n)), \tag{3.2}$$

and

$$r(n) = \sum_{i=1}^{q_{\mathrm{r}}} \alpha_{2,i}(n) r(n-i) + \sum_{i=1}^{q_{\mathrm{e}}} \beta_{3,i}(n) e_{\mathrm{i}}(n-i) + \vartheta_2(n), \quad \Sigma_{12}(n) = var(\vartheta_2(n)), \tag{3.3}$$

$$e_{\mathrm{i}}(n) = \sum_{i=1}^{q_{\mathrm{e}}} \beta_{2,i}(n) e_{\mathrm{i}}(n-i) + \sum_{i=1}^{q_{\mathrm{r}}} \alpha_{3,i}(n) r(n-i) + \upsilon_2(n), \quad \Sigma_{22}(n) = var(\upsilon_2(n)). \tag{3.4}$$

The time-varying strength of causality from r to e_{i} and from e_{i} to r are, respectively, defined as

$$F_{r \to e_{\mathrm{i}}}(n) = \ln \frac{\Sigma_{21}(n)}{\Sigma_{22}(n)}, \tag{3.5}$$

and

$$F_{e_{\mathrm{i}} \to r}(n) = \ln \frac{\Sigma_{11}(n)}{\Sigma_{12}(n)}. \tag{3.6}$$

If $F_{r \to e_{\mathrm{i}}}(n) > F_{e_{\mathrm{i}} \to r}(n)$, we can say that r Granger causes e_{i} at the time n, and vice versa. Generally, we need to set an appropriate threshold to determine whether the causal effect is significant or not. If an influential factor r Granger causes e_{i} in certain time intervals, we randomize the order of e_i such that the causality between r and e_i is annihilated. Note that the distribution of e_i remains unchanged during such a process. This is the so-called surrogate data approach. The shuffling procedure will be repeated for a number of times, say N_s, to yield a meaningful result. After all these processes, we then calculate the resulting $\kappa\%$ quantile for each time point, which is used to represent the threshold or the significant level of the Granger causality. Values above this level have a probability of occurring chance less than $1-\kappa\%$. A Granger causality relationship is considered to be significant when the causality strength surpasses the threshold. We note that it is common to set $\kappa\% = 95\%$ in the literature. We will illustrate the above procedure later in actual tests to be conducted for the DJIA.

3.2.1.2 Nonlinear Causality Test

For a given pair of time series $r(n)$ and $e_i(n)$, we let $e_i^m(n)$ be the m-length lead vector of e_i, $r^{Lr}(n-Lr)$ and $e_i^{Le}(n-Le)$ be the Lr-length and Le-length lag vectors of r and e_i, which are defined as

$$e_i^m(n) = [e_i(n), e_i(n+1), \cdots, e_i(n+m-1)], \quad n = 1, 2, \cdots, \tag{3.7}$$

$$r^{Lr}(n-Lr) = [r(n-Lr), r(n-Lr+1), \cdots, r(n-1)], \quad n = Lr+1, Lr+2, \cdots \tag{3.8}$$

and

$$e_i^{Le}(n-Le) = [e_i(n-Le), e_i(n-Le+1), \cdots, e_i(n-1)], \quad n = Le+1, Le+2, \cdots \tag{3.9}$$

respectively. The nonlinear Granger noncausality is defined in the context of conditional probability [7]. Given values of m, Le, and $Lr \geq 1$, for $\rho > 0$, r is said not to strictly Granger cause e_i if

$$\begin{aligned} &P\Big(\| e_i^m(n) - e_i^m(s) \| < \rho \mid \| e_i^{Le}(n-Le) - e_i^{Le}(s-Le) \| < \rho, \\ &\quad \| r^{Lr}(n-Lr) - r^{Lr}(s-Lr) \| < \rho \Big) \\ &= P\Big(\| e_i^m(n) - e_i^m(s) \| < \rho \mid \| e_i^{Le}(n-Le) - e_i^{Le}(s-Le) \| < \rho \Big), \end{aligned} \tag{3.10}$$

where $P(\cdot)$ and $\|\cdot\|$ denote probability and the maximum norm (supremum norm), respectively. The probability on the left-hand side of (3.10) is the conditional probability that two arbitrary m-length lead vectors of e_i are within a distance ρ of each other, given that the corresponding Le-length lag vectors of e_i and Lr-length lag vectors of r are within the same distance ρ of each other. We note that the condition on r is ignored on the right-hand side of (3.10).

Hiemstra and Jones [65] proposed an implementation of the above nonlinear Granger noncausality test by expressing the conditional probabilities in terms of the corresponding ratios of joint probability, in which (3.10) can be rewritten as

$$\frac{C_1(m+Le,Lr,\rho)}{C_2(Le,Lr,\rho)} = \frac{C_3(m+Le,\rho)}{C_4(Le,\rho)}, \tag{3.11}$$

where C_1, C_2, C_3 and C_4 are the correlation-integral estimators of the joint probabilities and are given as

$$\begin{aligned} C_1\,(m+Le,Lr,\rho) = P\Big(& \parallel e_{\mathrm{i}}^{m+Le}(n-Le) - e_{\mathrm{i}}^{m+Le}(s-Le) \parallel < \rho, \\ & \parallel r^{Lr}(n-Lr) - r^{Lr}(s-Lr) \parallel < \rho \Big), \end{aligned} \tag{3.12}$$

$$\begin{aligned} C_2\,(Le,Lr,\rho) = P\Big(& \parallel e_{\mathrm{i}}^{Le}(n-Le) - e_{\mathrm{i}}^{Le}(s-Le) \parallel < \rho, \\ & \parallel r^{Lr}(n-Lr) - r^{Lr}(s-Lr) \parallel < \rho \Big), \end{aligned} \tag{3.13}$$

$$C_3\,(m+Le,\rho) = P\Big(\parallel e_{\mathrm{i}}^{m+Le}(n-Le) - e_{\mathrm{i}}^{m+Le}(s-Le) \parallel < \rho \Big), \tag{3.14}$$

and

$$C_4\,(Le,\rho) = P\Big(\parallel e_{\mathrm{i}}^{Le}(n-Le) - e_{\mathrm{i}}^{Le}(s-Le) \parallel < \rho \Big). \tag{3.15}$$

For e_{i} and r with sampling size of N, the values of C_1, C_2, C_3 and C_4 can be easily estimated by the following numerical method:

$$\begin{aligned} C_{l_1}\,(L_1,L_2,\rho,n_c) = \frac{2}{n_c(n_c-1)} \sum\sum\nolimits_{n<s} & I\left(\parallel e_{\mathrm{i}}^{L_1}(n-Le) - e_{\mathrm{i}}^{L_1}(s-Le) \parallel < \rho \right) \\ & \cdot I\Big(\parallel r^{L_2}(n-Lr) - r^{L_2}(s-Lr) \parallel < \rho \Big), \end{aligned} \tag{3.16}$$

and

$$C_{l_2}(L_1,\rho,n_c) = \frac{2}{n_c(n_c-1)} \sum\sum\nolimits_{n<s} I\left(\parallel e_{\mathrm{i}}^{L_1}(n-Le) - e_{\mathrm{i}}^{L_1}(s-Le) \parallel < \rho \right), \tag{3.17}$$

where $l_1 = 1,2$; $l_2 = 3,4$; L_1 and L_2 represent their first two parameters; $n,s = \max(Le,Lr)+1,\cdots,N-m+1$; $n_c = N+1-m-\max(Le,Lr)$; $I(\cdot)$ equals 1 when the condition inside $(\cdot)$ is fulfilled and 0 otherwise.

Given values of m, Le and $Lr \geq 1$, for $\rho > 0$, the criterion of noncausality is that if r does not strictly Granger cause e_{i}, then

$$\sqrt{n_c}\left[\frac{C_1(m+Le,Lr,\rho,n_c)}{C_2(Le,Lr,\rho,n_c)} - \frac{C_3(m+Le,\rho,n_c)}{C_4(Le,\rho,n_c)} \right] \to N\left(0,\sigma^2(m,Le,Lr,\rho)\right). \tag{3.18}$$

The estimation of $\sigma^2(m,Le,Lr,\rho)$ can be found in [65].

In our framework, the nonlinear causality test is applied to series $\Sigma_{12}(n)$ of (3.3) and $\Sigma_{22}(n)$ of (3.4), which are similar to the so-called VAR residuals reported in [65]. For easy reference and for future use, we denote CS and TVAL, respectively, as the differences between the left- and right-hand sides of (3.11) and the standardized test statistic in (3.18). We note that there is no standard procedure in selecting the parameters Le, Lr and ρ. We follow the Monte Carlo approach given in [64] to standardize all series involved before conducting causality tests and set $\rho = 1.5\sigma$, where $\sigma = 1$ is the standard deviation of the standardized time series. We also set the lead length to be 1, i.e., $m = 1$, and the lag lengths $Le = Lr$, using common lag lengths from 1 to 10.

3.2.2 Redundant Variable Test

Before finalizing the candidates for the input of our framework, we might need to eliminate the possible multicollinearity among the selected influential factors, i.e., to remove some of them, which are redundant. A high degree of multicollinearity might produce invalid results in predicting the market. Our adaptive filter is a time-varying linear model, thus the existence of multicollinearity of the input channels may cause the time-varying parameters to become unstable. To avoid such a problem, for each influential factor that is regressed with the others, a tolerance is calculated as $1 - R^2$, where R^2 is the coefficient of determination [121] and is given as

$$R^2 = 1 - \frac{\Sigma(y - y_\mathrm{h})^2}{\Sigma(y - \bar{y})^2}, \tag{3.19}$$

where y is the actual output, y_h is the predicted output and $\bar{y}$ is the mean of y. A tolerance close to 1 means there is little multicollinearity, whereas a small value of tolerance suggests that multicollinearity should be noted.

3.3 Influential Factors of Dow Jones Industrial Average

3.3.1 Empirical Selection

Benefiting from its maturity, various derivatives are available for the U.S. stock market, providing a valuable source of sentiment indicators. Sun [122] provided an empirical analysis on a wide range of indicators, based on which five leading economic and sentiment indicators are selected to investigate the contributions among them in generating the external market force. Of the five selected factors, three are economic indicators, i.e., the interest rate indicator (IRI), the oil price (OP) and the Baltic dry index (BDI); and two are sentiment indicators, i.e., the Chicago Board Options Exchange DJIA volatility index (VXD) and the exchange rate of the Euro against the Japanese Yen (EUR/JPY).

1. *Interest Rate Indicator (IRI)*
   For the economic indicators, the interest rate is one of the most well-known factors that have a direct impact on the general trend of the stock market. The size

of the federal funds rate target (FFRT) changes is often applied to investigate the role that the interest rate plays in the U.S. stock market [60, 125]. Since the federal funds future rate (FFFR) embodies market expectation on the monthly average of the daily effective funds rate, it is frequently used in measuring market reactions to interest rate changes [15, 80]. We construct a similar indicator that takes market expectation on a daily basis into consideration. Since the daily effective federal funds rate (DEFFR)[1] is a volume-weighted average of rates on trades arranged by a group of federal funds brokers who report to the Federal Reserve Bank of New York each day, we use the difference between DEFFR and FFRT, which is defined as the interest rate indicator (IRI), as a gauge of the influence of interest rates on the U.S. stock market:

$$\text{IRI} = \text{DEFFR} - \text{FFRT}. \tag{3.20}$$

2. *Oil Price (OP)*
   From the commodity sector, we select changes in the oil price (OP) as another key economic indicator. The oil price is using the Cushing West Texas Intermediate (WTI) spot price, obtained from Energy Information Administration. Its relative change $c(n)$ is calculated by

$$c(n) = 100 \times \frac{o(n) - o(n-1)}{o(n-1)}, \tag{3.21}$$

   where $o(n)$ is the value of oil price.

3. *Baltic Dry Index (BDI)*
   The third economic indicator is the Baltic Dry Index (BDI), a daily average of global shipping prices for dry bulk cargoes. Functioning as an assessment of global trade and free of manipulation and speculation, it acts as an excellent leading indicator of economic activities. We also use its changes calculated in the same fashion as that in (3.21) to measure its influence on the stock market.

4. *Chicago Board Options Exchange DJIA Volatility Index (VXD)*
   Two sentiment indicators chosen for the DJIA are the VXD and the EUR/JPY currency pair. The VXD is a futures contract based on the prices of options on the DJIA traded at the Chicago Board Options Exchange (CBOE). It is a kind of stock fear index, reflecting the market expectation of the DJIA volatility over the next 30 days. Again, the relative changes of the VXD and the EUR/JPY exchange rate, calculated in the similar way as that in (3.21), are used in our process.

5. *Exchange rate of the Euro against the Japanese Yen (EUR/JPY)*
   Financial crisis would inevitably disseminate fear, resulting in investors being averse to risk. The currency pair EUR/JPY is a good choice for such an indicator. In the work of Sun [122], the empirical comparison of the DJIA and the EUR/JPY

[1] The definition is from the Federal Reserve Bank of New York.

exchange rate suggested that they nourish each other through interactions. Due to this reason, we choose a currency pair that does not include the U.S. dollar.

Before conducting tests, we need to synchronize the data of the indicators and the stock prices. Specifically, if an indicator has no data available on certain samples, on which the stock market still traded, we will use the value of the indicator from the immediately preceding sample. On the other hand, if the stock price has no data available on a certain sample, but the indicator does, the latter will be removed accordingly. After synchronization, we next perform data normalization for all the series except the IRI using

$$\tilde{v}(n) = \frac{v(n) - \bar{v}}{s_v}, \tag{3.22}$$

where $v(n)$ denotes the original series, $\tilde{v}(n)$ denotes the normalized series, $\bar{v}$ and s_v are, respectively, the mean and the standard deviation of v. Figures 3.1 to 3.5 show the synchronized and normalized data of the selected input indicators for the DJIA over the period of interest.

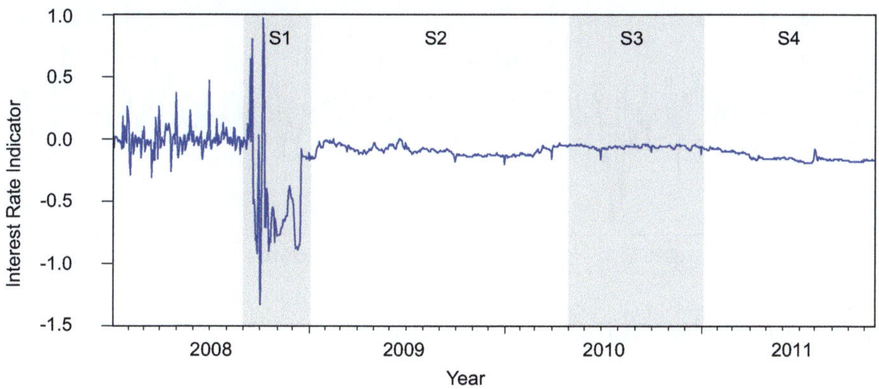

Fig. 3.1 Interest rate indicator (IRI) for the U.S. stock market

3.3.2 *Time-Varying Causality Test Results*

When the lag length of the internal residue of the DJIA (see Figure 2.6) reaches 4, the corresponding Durbin-Watson statistic is 1.98 and its p value of Breusch-Godfrey test is 0.43, both showing no autocorrelation in the residuals. We thus set the lag length of the series of e_{i} to 4. To evaluate the effectiveness of indicators within a half-month period over the internal residue, the lag lengths of all input indicators are set to 10. The resulting time-varying causal relationships between each of the five selected indicators and the internal residue of the DJIA are shown in Figures 3.6 to 3.10. For each pair, we can observe that the causality strength from the indicator to the internal residue is higher than that in the opposite direction. Threshold values are then calculated by the surrogate data approach. We shuffle the series

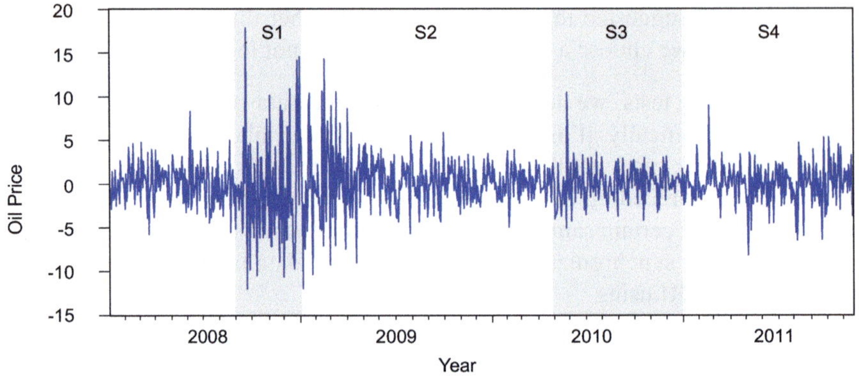

Fig. 3.2 Oil price (OP) for the U.S. stock market

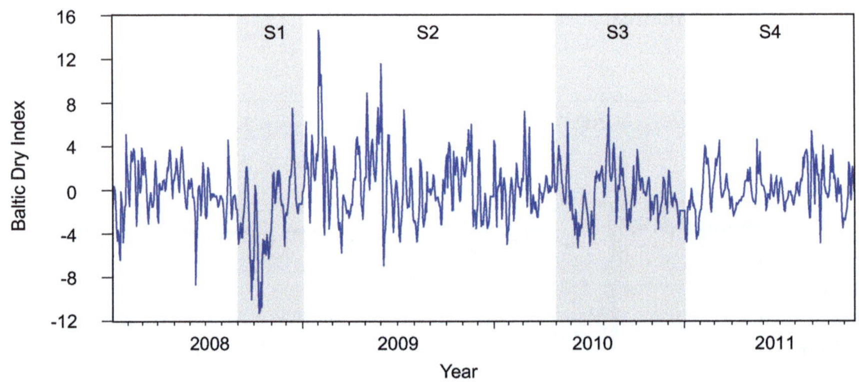

Fig. 3.3 Baltic dry index (BDI) for the U.S. stock market

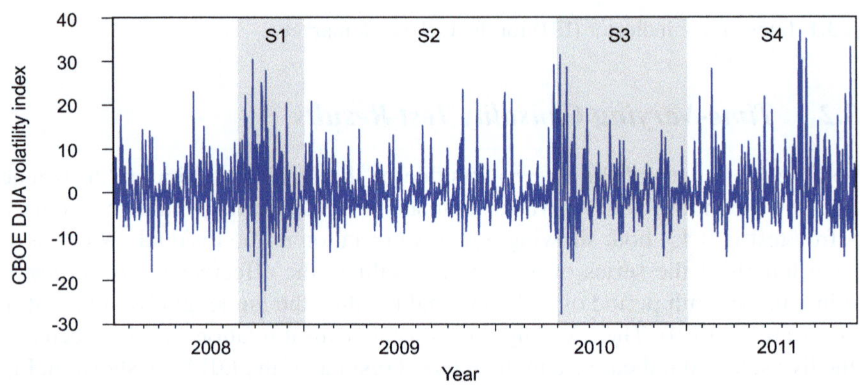

Fig. 3.4 CBOE DJIA volatility index (VXD) for the U.S. stock market

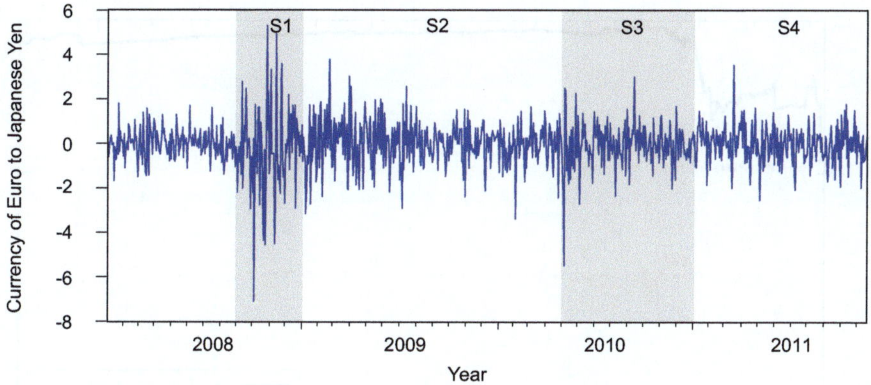

Fig. 3.5 Currency of Euro to Japanese Yen (EUR/JPY) for the U.S. stock market

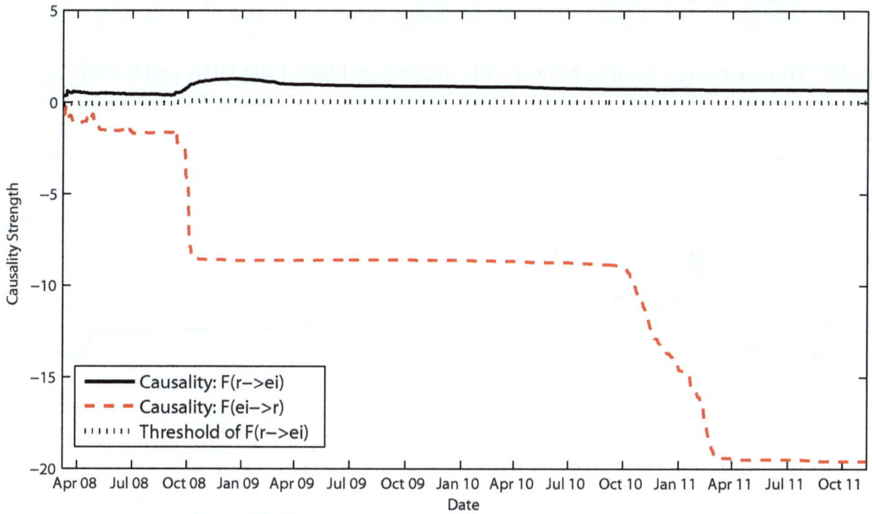

Fig. 3.6 Time-varying causality between the internal residue of the DJIA and the IRI

of e_i for 200 times, i.e., $N_s = 200$, and set the quantile $\kappa\%$ as 95% in calculating the threshold of the causality strength. As shown in our testing results, four out of five indicators (except the pair associated with the BDI) significantly Granger cause the internal residue as their causality strengths exceed the corresponding thresholds over the entire sampling period.

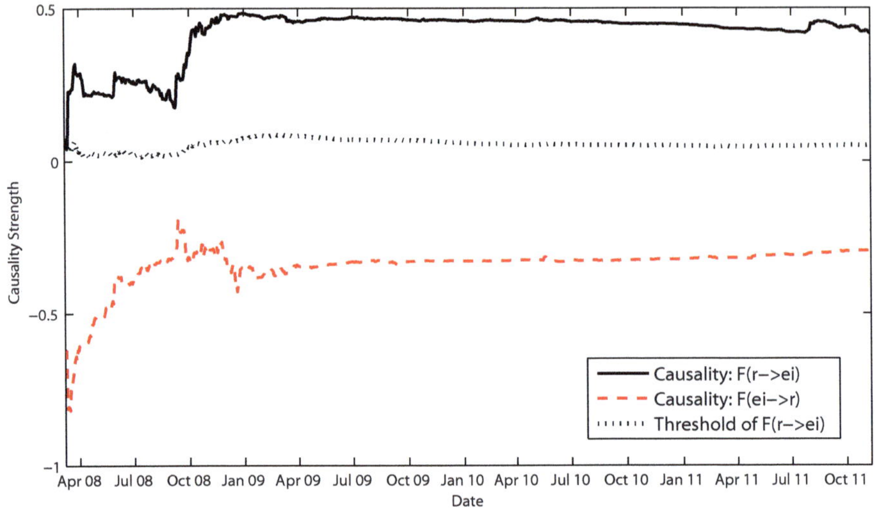

Fig. 3.7 Time-varying causality between the internal residue of the DJIA and the OP

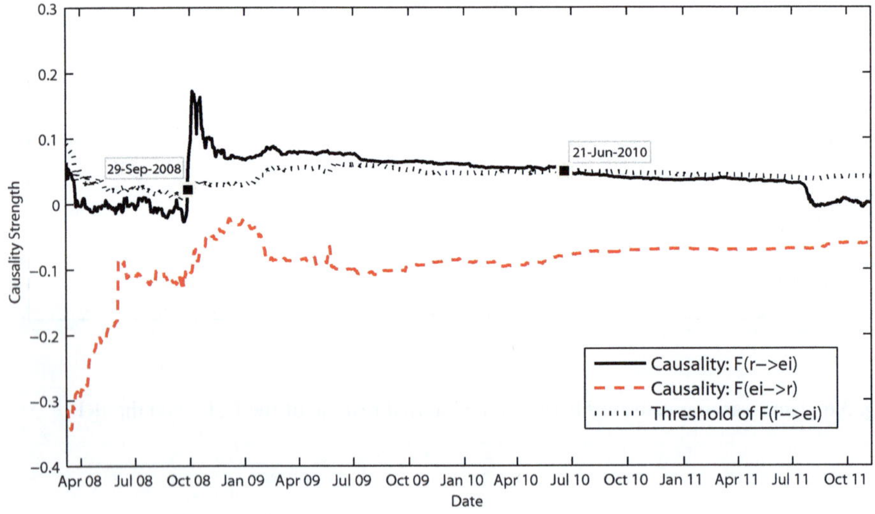

Fig. 3.8 Time-varying causality between the internal residue of the DJIA and the BDI

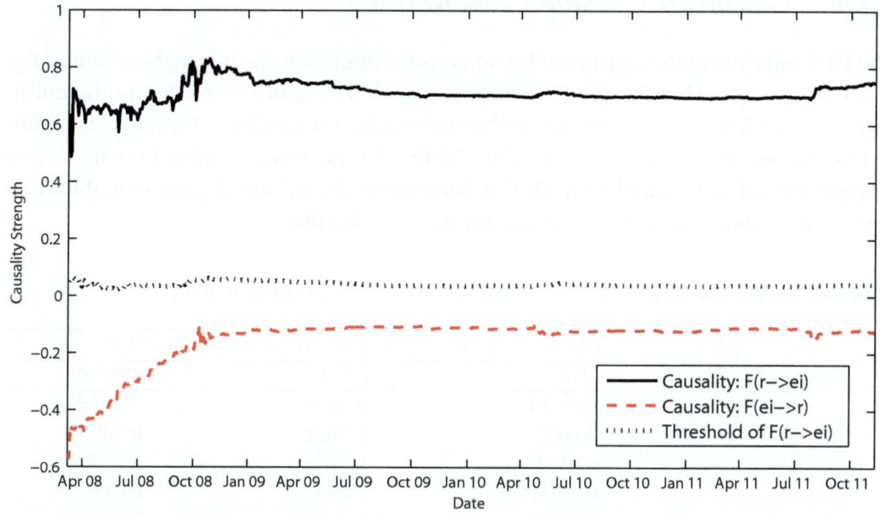

Fig. 3.9 Time-varying causality between the internal residue of the DJIA and the VXD

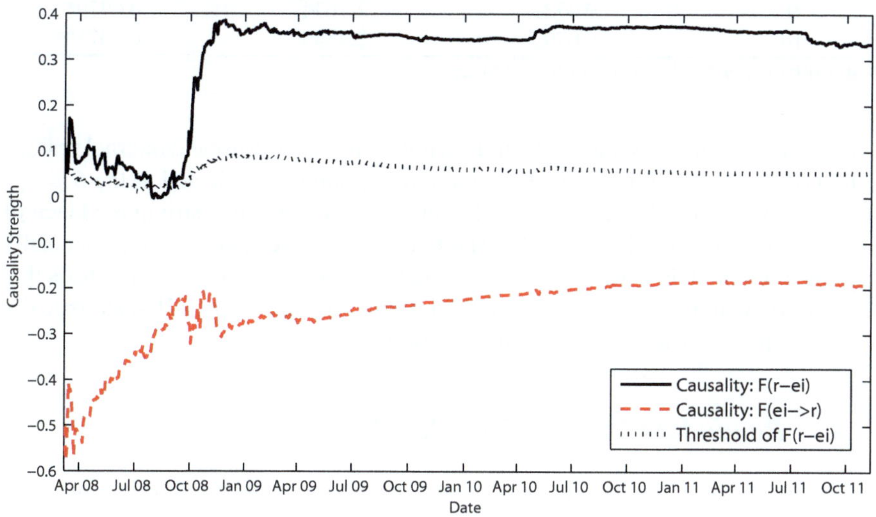

Fig. 3.10 Time-varying causality between the internal residue of the DJIA and the EUR/JPY

3.3.3 Nonlinear Causality Test Results

The BDI only presents significant linear causal effect over the internal residue of the DJIA from September 29, 2008 to July 21, 2010. Therefore, considering the entire period as a whole, it is necessary to further conduct a nonlinear Granger causality test on the pair associated with the BDI. Table 3.1 reports the results of the nonlinear Granger causality test applied to BDI-related series $\Sigma_{12}(n)$ and $\Sigma_{22}(n)$ with the same lag lengths. The p value is calculated for a one-sided test.

Table 3.1 Nonlinear Granger causality test results in the U.S. stock market

	BDI does not cause the internal residue of the DJIA		
$Le = Lr$	CS	TVAL	p value
1	9.1884×10^{-5}	0.0435	0.4826
2	0.0040	1.0691	0.1425
3	0.0116	1.7759	0.0379**
4	0.0200	2.0952	0.0181**
5	0.0300	2.3977	0.0082**
6	0.0467	2.9158	0.0018**
7	0.0625	3.0650	0.0011**
8	0.0690	2.8864	0.0019**
9	0.0731	2.7644	0.0029**
10	0.0790	2.4545	0.0071**

** Significance at 5% level for a one-sided test.

The null hypothesis is that the BDI does not nonlinearly cause the internal residue of the DJIA. As shown in Table 3.1, this null hypothesis is rejected at the 5% significance level when the lag length is larger than two. This is a strong evidence of nonlinear Granger causality from the BDI to the internal residue of DJIA, especially when the lag length is long. As such, we conclude from previous causality tests that these five indicators all Granger cause the internal residue of the DJIA, statistically supporting the rationality of our input selection.

3.3.4 Redundant Variable Test Results

Finally, we perform multicollinearity tests among the indicators. The results are shown in Table 3.2. We note that the EUR/JPY exchange rate gives a relatively small tolerance, and thus could be removed. The IRI, OP, BDI and VXD indicators are finally used as the input to our system adaptation framework to forecast the DJIA.

Table 3.2 Multicollinearity test results in the U.S. stock market

	IRI	OP	BDI	VXD	EUR/JPY
Tolerance	0.971	0.820	0.992	0.785	0.626

Chapter 4
Analysis of Dow Jones Industrial Average

4.1 Introduction

We present in this chapter the detailed analysis of the Dow Jones Industrial Average (DJIA) index using our system adaptation framework together with the influential factors selected in Chapter 3. The U.S. stock market is remarkably important not only because the U.S. national economy is the largest in the world, but also because it has great influence on other markets. Generally, global stock markets respond quickly and follow closely to the trend of the U.S. market, especially in abnormal situations when the market is highly volatile. On the other hand, most economic theories and assumptions are proposed based on the research of the developed financial systems, which have larger and more liquid stock markets, more mature economies and more effective financial regulatory systems. As a typical representative of the developed markets, the U.S. market is also the most favorable object used to prove empirical or theoretical propositions in the academic research. The U.S. stock market and other U.S. capital markets work together in a highly efficient way to compose an essential part of the U.S. financial system. Furthermore, the U.S. has a well-established statistical system and one of the best financial regulatory systems. Its financial data are complete, accurate, reliable and most openly accessible. All these characteristics make the U.S. market a great example for research.

The DJIA index, the Standard & Poor's 500 Index (S&P 500) and the Nasdaq Composite Index (NASDAQ) are three major U.S. stock indices with the DJIA being the most renowned among them. It is an indicator for the health of the American economy. In the global market, the importance of the DJIA is further acknowledged beyond its domestic roles as it is one of the most important economic indices in the world. The thirty companies listed in the DJIA components, such as the Bank of America, Coca-Cola, General Electric, Microsoft, to name a few, are all well-known multinational enterprises. They cover a variety of big industries, and their performances are in the wake of the global economy. We thus use the DJIA as the representative of the U.S. stock market for illustrating the effectiveness and performance of the proposed system adaptation framework.

X. Zheng & B.M. Chen: *Stock Market Modeling and Forecasting*, LNCIS 442, pp. 43–51.
DOI: 10.1007/978-1-4471-5155-5_4

Other than the DJIA, the S&P 500 is also a commonly adopted benchmark of the U.S. stock market. The S&P 500 includes 500 companies on the large-cap sector of the market. It provides broader market coverage and is weighted by the company's market capitalization instead of the stock price as used in the DJIA. The S&P 500 will be adopted to represent the influence of the U.S. market on others, such as the stock markets in Hong Kong and Singapore, in the next chapter. On the other hand, although the NASDAQ is highly followed by investors who are interested in technology stocks, it is more volatile than the other two, since it includes many small but high-growth stocks. It is currently not used in our study.

For easy references, we recall that for the period under investigation for the DJIA, i.e., from January 2008 to November 2011, a corresponding internal OE model under the system adaptation framework has been obtained in Chapter 2 using the historical closing prices from January 2003 to December 2007. The resulting OE model is given as

$$H_{\mathrm{DJIA}}(z) = \begin{bmatrix} \dfrac{0.9574z^{-1} - 0.5034z^{-2} - 0.321z^{-3} + 0.565z^{-4}}{1 - 1.329z^{-1} + 0.7312z^{-2}} \\ \dfrac{-5.127z^{-1} + 2.086z^{-2} + 0.3914z^{-3} + 0.8506z^{-4}}{1 - 0.62z^{-1} - 0.1239z^{-2}} \\ \dfrac{2.417z^{-1} + 3.049z^{-2} - 2.298z^{-3} - 1.82z^{-4}}{1 - 0.3215z^{-1} - 0.4643z^{-2}} \end{bmatrix}^{\mathrm{T}}. \tag{4.1}$$

We have also identified in Chapter 3 four crucial economic and sentiment indicators for the DJIA, i.e., the interest rate, the oil price, the Baltic dry index and the Chicago Board Options Exchange DJIA volatility index, as the input to the framework, acting as the source of the external force. Having the internal model and the input force, an adaptive filter can then be estimated to complete the overall structure of the proposed system adaptation framework.

In what follows, we first introduce the variables to measure the predicting performances, and then explain the parameter selection and the whole one-step-ahead prediction process together with a preliminary analysis to assess the influence of each influential factor in the market prediction as well as the overall analysis of the results obtained in this study.

4.2 Measure of Predicting Performance

We adopt the adjusted R^2, mean absolute error (MAE) and root mean squared error (RMSE) to measure the predicting performance of our framework. The original R^2, i.e., the coefficient of determination of (3.19), is to provide the proportion of variability in a data set that is accounted for by a statistical model [121]. The adjusted R^2, i.e., $\bar{R}^2$, is a slight modification of R^2 by adjusting for the sample size and degree of freedom, which is defined as

$$\bar{R}^2 = 1 - \frac{K-1}{K-w-1}\frac{\Sigma(y-y_{\mathrm{h}})^2}{\Sigma(y-\bar{y})^2}, \tag{4.2}$$

where K is the sample size, w is the total number of regressors in the model, y is the actual output, y_{h} is the predicted output and $\bar{y}$ is the mean of y. This statistic increases only if the new term improves the model more than what would be expected by chance; therefore, we use it to measure the contribution of each indicator to the internal residue. The MAE and RMSE are respectively defined as

$$\mathrm{MAE} = \frac{1}{K}\sum_{n=1}^{K}|e(n)|, \tag{4.3}$$

and

$$\mathrm{RMSE} = \left[\frac{1}{K}\sum_{n=1}^{K}e^2(n)\right]^{\frac{1}{2}}, \tag{4.4}$$

where K is the number of samples interested.

4.3 Adaptive Filter and Predicting Performance

As mentioned earlier, the autoregressive (AR) model is used to determine the lag length of $e_{\mathrm{i}}(n)$ in model (2.22). For the data of the internal residue, when the lag length reaches 4, the Durbin-Watson statistic is 1.98 and the p value of Breusch-Godfrey test is 0.43, both showing no autocorrelation in the residuals. Thus, we set the corresponding n_{a} to be 4. To evaluate the effect of the selected indicators within half a month (about 10 trading days) over the internal residue $e_{\mathrm{i}}(n)$, the lag lengths of all the input indicators are set to be 10. An interesting observation shows that when the time delay of the IRI starts from third order, the model yields the best results. As such, we set the lag length of the IRI in our test to be from 3 to 12. To be practical, only the data from January to August 2008 are used when performing the hyperparameter optimization. The influence of each indicator is investigated by using the indicator data and its AR components alone to predict the internal residue, before combining all selected indicators together as the input to test the prediction ability of the whole framework.

4.3.1 *Preliminary Analysis*

Following the procedure given in Section 2.4 of Chapter 2, we proceed to estimate the hyperparameters for each influential indicator. Initially, we set all the hyperparameters as 0.002 and the covariance matrix $\hat{P}$ as a diagonal matrix with all its diagonal entries equal to 10^5. Note that the same initial values will be adopted when combining all indicators together as the input for the prediction in the next section.

In this part, we have 14 state variables and 14 hyperparameters in the Q_r matrix. With the estimated hyperparameters, one-step-ahead prediction results can be obtained. Table 4.1 summarizes the associated adjusted R^2 ($\bar{R}^2$).

Table 4.1 Prediction results ($\bar{R}^2$) for the U.S. stock market

	Subperiod S1	Subperiod S2	Subperiod S3	Subperiod S4
AR	21.48%	16.05%	20.09%	22.08%
AR and IRI	72.86%	18.95%	18.15%	34.92%
AR and OP	45.96%	44.88%	44.85%	44.61%
AR and BDI	21.83%	27.75%	20.55%	23.83%
AR and VXD	53.07%	48.43%	49.08%	64.64%

Generally, the force variance is also considered as the market variance. In the U.S. stock market, the market variance in Subperiod S1 can be better explained than in Subperiod S2. Among the selected indicators, the IRI is the most dominant factor in Subperiod S1 but contributes much less to the internal residue after that. The VXD is another significant indicator in all time intervals, which indicates that the investors' sentiment measured by the VXD plays an important role. The OP has also shown its importance as its $\bar{R}^2$ are close to the VXD. We note that even though we categorize the IRI as the economic indicator, it contains sentiment elements by including the DEFFR, a market data reflecting the public expectations.

From the beginning of Subperiod S1, the U.S. financial crisis was entering into a dangerous phase. It was accompanied by a series of collapses of financial institutions, such as the takeover of Fannie Mae and Freddie Mac, the bankruptcy of Lehman Brothers and the merge of Bank of America and Merrill Lynch. The devastating effects they triggered began to spread to all other economic sectors, undermining the confidence of investors. An unprecedented $700 billion rescue plan was enacted by the U.S. government in October 2008. However, the original plan was rejected by the House of Representatives on September 29, 2008. All these events resulted in a drastic instability in the U.S. stock market, which is shown in the market internal residue from September to December 2008 (see Figure 2.6). It is also reflected in the expectations of investors on the market quantified by the IRI. Since the U.S. Federal Reserve decided to keep the FFRT at 2% on September 16, 2008, it deviated from the major expectation on the open market. As highlighted by the first shaded area in Figure 3.1, although the FFRT was consecutively cut for two times in October 2008, the IRI still underwent a drastic fluctuation.

Interestingly, the IRI seems to be less important in Subperiod S2. A possible explanation could be due to the fact that Federal Reserve slashed the FFRT to 0.25% in December 2008 and keeps it low till today. That makes the IRI stay in a rational range as it matches market expectations. In Subperiod S2, the OP seriously affected the U.S. stock market, accounting for nearly 50% of the market variance. Unlike the interest rate, the relationship between the OP and the stock market is complicated and in debate in the academic and investment circles. As supported by our testing results, the OP led the stock market and its influence exceeded much more than that of the IRI after 2009. From the beginning of 2009, the OP began to rebound but was still rational. The DJIA moved in the same directions with it as a follower. The strengthening of oil demand can be considered as a sign of the consolidation of economic recovery in the U.S. With more money and credit poured in the economy, consumer spending increased, and the stock market reacted to these information and rallied. It is not surprising that the article of McKay entitled "Oil Lifts Dow to 10062.94 in Late Rally," made the headline of the Wall Street Journal on October 16, 2009 (see [96]) on the analysis of this situation.

4.3.2 Adaptive Filter Hyperparameters

Combining together all the selected influential factors, we once again estimate the hyperparameters of the corresponding adaptive filter with the complete set of the input and output market data from January to August 2008. As in the previous subsection, the initial values of the hyperparameters and the covariances are set as 0.002 and 10^5, respectively. We obtain the estimation of the hyperparameters of the corresponding adaptive filter as follows:

1. A diagonal matrix Q_r with its main diagonal entries being given as

$$0.0025,\ 0.0032,\ 1.1226\times10^{-4},\ 2.3812\times10^{-4}, \tag{4.5}$$

which are corresponding to the estimations of $\hat{a}_1, \hat{a}_2, \hat{a}_3, \hat{a}_4$.

2. The estimations of $\hat{b}_{1,1}, \hat{b}_{1,2}, \cdots, \hat{b}_{1,10}$, are given as

$$0.0100,\ 0.0211,\ 0.0079,\ 0.0075,\ 0.0380, \tag{4.6}$$

$$0.7546,\ 6.7498\times10^{-4},\ 0.3482,\ 0.1456,\ 0.1433, \tag{4.7}$$

the coefficients associated with the Interest Rate Indicator (IRI) input.

3. The estimations of $\hat{b}_{2,1}, \hat{b}_{2,2}, \cdots, \hat{b}_{2,10}$, are

$$1.7388\times10^{-4},\ 2.6124\times10^{-4},\ 0.0056,\ 0.0656,\ 3.0712\times10^{-4}, \tag{4.8}$$

$$1.9431\times10^{-4},\ 0.0011,\ 7.1140\times10^{-4},\ 0.0584,\ 0.0141, \tag{4.9}$$

the coefficients associated with the Oil Price (OP) input.

4. The estimations of $\hat{b}_{3,1}, \hat{b}_{3,2}, \cdots, \hat{b}_{3,10}$, are

$$1.7249\times10^{-4},\ 0.0153,\ 0.0015,\ 0.2589,\ 0.0023 \tag{4.10}$$

$$5.3271\times10^{-4},\ 7.9624\times10^{-4},\ 2.9077\times10^{-4},\ 4.3319\times10^{-4},\ 0.0267, \tag{4.11}$$

the coefficients associated with the Baltic Dry Index (BDI) input.

5. Finally, the estimation of $\hat{b}_{4,1}, \hat{b}_{4,2}, \cdots, \hat{b}_{4,10}$, are

$$1.9714\times 10^{-4},\ 5.9100\times 10^{-4},\ 7.3694\times 10^{-4},\ 3.5652\times 10^{-4},\ 1.9268\times 10^{-4}, \quad (4.12)$$

$$0.0012,\ 2.5038\times 10^{-8},\ 5.3246\times 10^{-4},\ 3.8426\times 10^{-4},\ 1.1683\times 10^{-4}, \quad (4.13)$$

the coefficients associated with the VXD input.

With the identified Q_{r} and the initial choice of $\hat{X}(0) = 0$, $\hat{P}(0)$ being a diagonal matrix with all its diagonal entries equal to 10^5, and $\hat{H}(0)$ being set to its corresponding input data, the one-day-ahead prediction results can be obtained through the iterative process as given in (2.38) to (2.41).

4.3.3 *Predicting Performance*

We compare the predicting ability of our framework with the commonly adopted autoregressive moving average model with exogenous input (ARMAX). The lag lengths for the ARMAX model are set similarly as those used in our framework, i.e., 4 for the AR and MA (moving average) terms and 10 for all the exogenous inputs. Summarized in Table 4.2 are the prediction error results of our proposed framework and the ARMAX method. It is obvious that the system adaptation framework greatly outperforms the ARMAX approach, especially in Subperiods S1. Figures 4.1 and 4.2 are the detailed prediction results of two time frames from September 2008 to January 2009 and from May 2010 to December 2010, respectively. The effectiveness of the system adaptation framework structure, the ability of the dynamical design of the adaptive filter and the distinguished function of the internal model are comprehensively testified.

It is clear that the predicting or forecasting ability of the system adaptation framework is far superior especially for complicated economic situations when the market is highly volatile. For example, in Subperiod S1, our framework gives amazingly accurate prediction results, whereas the ARMAX model totally fails to measure the dynamics of the market. Nevertheless, we should also note that under the framework structure, the predictions in Subperiods S1 and S2 are better than those in Subperiods S3 and S4. Such a phenomenon could probably be due to the weakening of the determinant effect of the input sources. For references, we note that the average daily changes of the closing prices of the DJIA from Subperiod S1 to S4 are 263.82, 85.61, 79.29 and 110.19, respectively. Compared with the MAE generated by the ARMAX which are slightly within the range of this average daily price changes, our framework provides much smaller and reasonable MAE in all subperiods. Hence, our framework is much more meaningful.

Finally, we would like to emphasize once more that our framework is structured in a systematic and flexible fashion. It can be easily expanded to incorporate more market information and to capture more market dynamics. For example, how to structure the framework and to select appropriate influential factors that are suitable

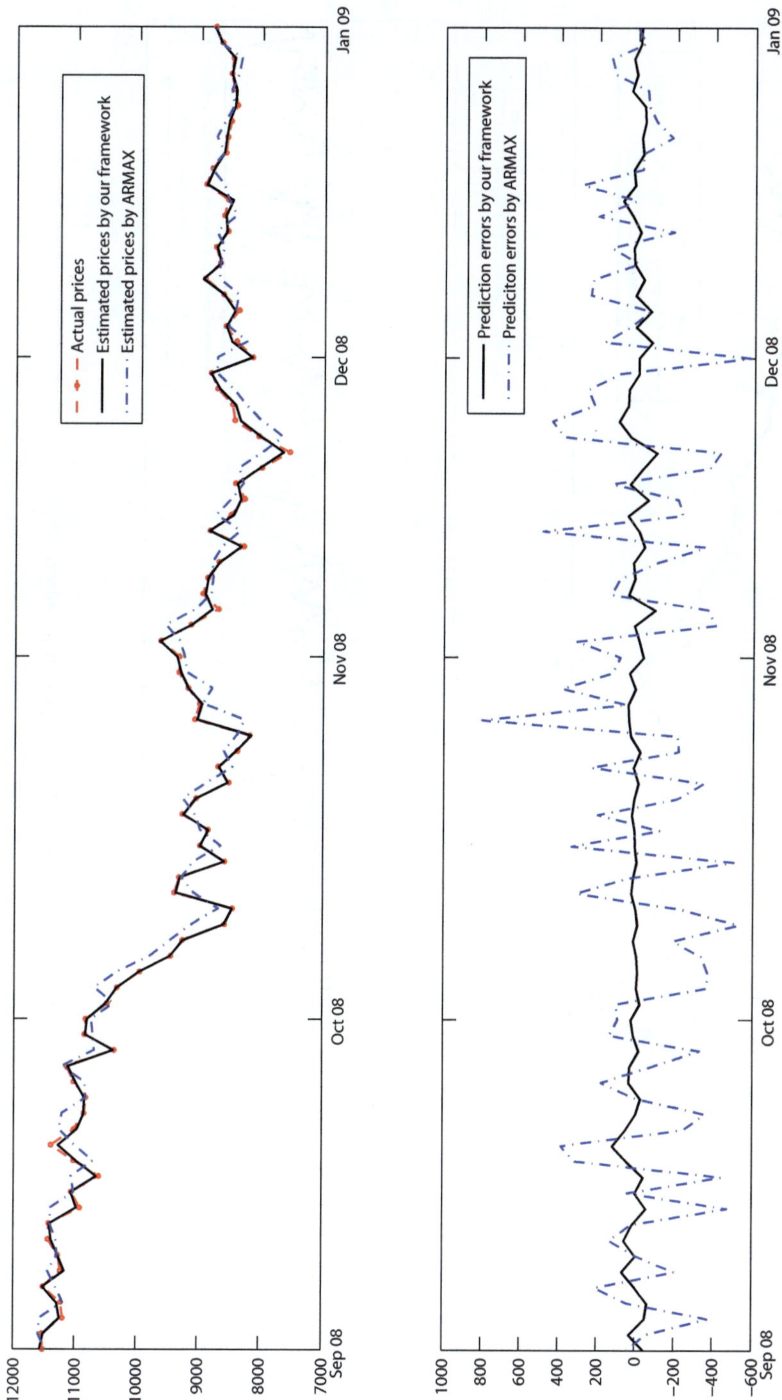

Fig. 4.1 Prediction results of the proposed framework and the ARMAX approach in Subperiod 1

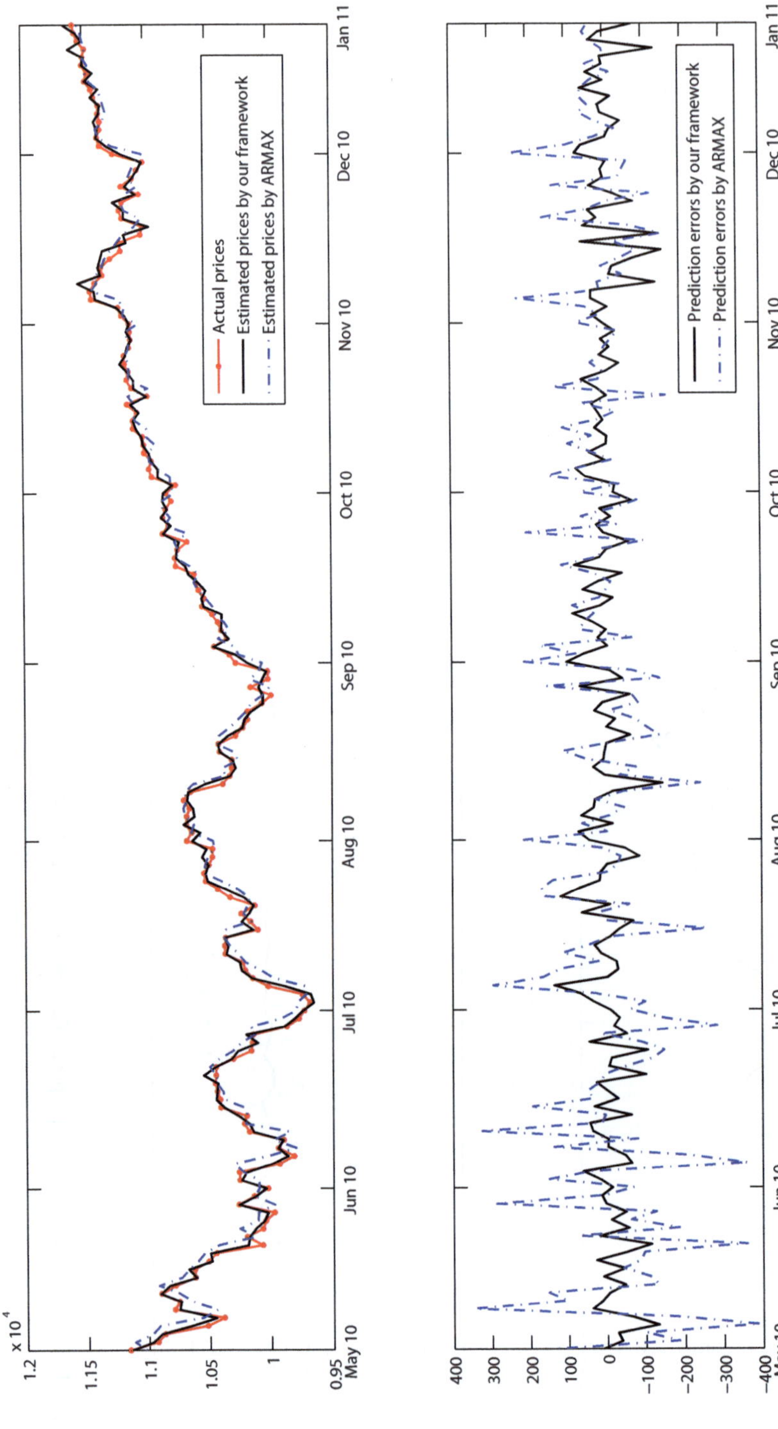

Fig. 4.2 Prediction results of the proposed framework and the ARMAX approach in Subperiod 3

Table 4.2 Comparison of the prediction results between the ARMAX and the system adaptation framework approaches for the DJIA index

	Subperiod S1 Sep. 2008 – Dec. 2008		Subperiod S2 Jan. 2009 – Apr. 2010	
	MAE	RMSE	MAE	RMSE
ARMAX	215.61	272.47	88.09	115.96
System adaptation framework	29.82	38.69	29.81	39.50
Improvement ((%))	86.17	85.80	66.16	65.94

	Subperiod S3 May 2010 – Dec. 2010		Subperiod S4 Jan. 2011 – Nov. 2011	
	MAE	RMSE	MAE	RMSE
ARMAX	85.60	119.27	110.15	147.87
System adaptation framework	36.18	48.77	35.48	45.83
Improvement ((%))	57.73	59.11	67.79	69.01

for high frequency trading could be an interesting direction for future research. Nevertheless, we will summarize our observations and possible future research topics later in the concluding chapter, i.e., Chapter 8.

Chapter 5
Selected Asian Markets

5.1 Introduction

Over the last two decades, one had witnessed the increasing importance and global impact of the Asian economy. It is believed that the share of the world economic growth held by the Asia-Pacific region is likely to keep advancing, enhancing Asia's role as a world economic engine in the future [70]. Besides some developed markets like Japan, Hong Kong and Singapore, several important emerging markets in developing countries, where increasingly open economy and high growth expectations are presented, are worth studying. The emerging market provides investors a great opportunity of profit. Although this kind of investment comes with high risk, the potential of high return still attracts a large amount of foreign and domestic investment. Among all the emerging markets, China is an excellent representative because of its rapid growth in economy. The stock market in China has expanded tremendously for the last couple of decades. It has unique features due to immature financial and statistical systems in China. All these properties constitute its unique research value.

To further examine the effectiveness of our framework, we proceed in this chapter to investigate the stock markets in China, Hong Kong and Singapore. More specifically, we will study the characteristics and influential factors of the Shanghai Stock Exchange (SSE) Composite Index of China, the Hang Seng Index (HSI) of Hong Kong and the Straits Times Index (STI) of Singapore, respectively. We note that Hong Kong and Singapore are the most important offshore finance markets in Asia, both of which have well-developed infrastructures. The Shanghai market, however, has exhibited some very unique features. Some influential factors, such as the interest rate indicator, seem to be not so effective for the Chinese market.

5.2 Shanghai Stock Exchange Composite Index

The Shanghai Stock Exchange (SSE) is the largest market in China and the third largest in the world by market capitalization. However, it is still not entirely open to

X. Zheng & B.M. Chen: *Stock Market Modeling and Forecasting*, LNCIS 442, pp. 53–89.
DOI: 10.1007/978-1-4471-5155-5_5

foreign investors and is significantly influenced by the Chinese government. Many characteristics of the China stock market are very unique. In this section, we once again focus on the impact of the 2007 global financial crisis on the China market, in which the SSE Composite Index plummeted more than 70% in one year starting from October 2007. We take the daily closing prices of the SSE index from the beginning of 2007 to November 2011, including the rapid rise and crash phases, for examination. The influential factors of the China and U.S. markets are also compared to find differences between the developed market and the emerging market.

The internal residue of the SSE is obtained through the same procedure as that for the DJIA studied in previous chapters. The internal OE model for the SSE under the system adaptation framework is estimated based on its daily closing prices from 2002 to 2006, and the resulting OE model is given by

$$H_{\mathrm{SSE}}(z) = \begin{bmatrix} \dfrac{0.5799z^{-1}-0.7365z^{-2}+0.506z^{-3}-0.2967z^{-4}}{1-0.7929z^{-1}-0.1169z^{-2}} \\ \dfrac{-2.234z^{-1}+1.558z^{-2}-1.158z^{-3}-0.3087z^{-4}}{1-0.8926z^{-1}+0.6835z^{-2}} \\ \dfrac{2.646z^{-1}-1.092z^{-2}-1.061z^{-3}-0.1253z^{-4}}{1-0.6328z^{-1}-0.2496z^{-2}} \end{bmatrix}^{\mathrm{T}}. \tag{5.1}$$

5.2.1 *Input Selection*

In what follows, we proceed to identify potential economic and sentiment influential factors for the SSE. Various influential factors of the China stock market have been studied in the literature. For example, Zheng and Wong [142] adopted a two-stage bivariate GARCH model to analyze the conditional dependence between the so-called A-type and B-type Shares in the China stock market, and the impacts of the U.S. and Hong Kong markets over the Chinese counterpart. Yao *et al.* [135] analyzed the relationships between the SSE and ten banking stocks listed in the market. The indicators related to the interest rate, money supply and inflation have also been investigated and reported.

Generally, there is a lack of sentiment indicators for the China markets. Since China's derivatives market is still relatively small and some market data are manipulated by the Chinese government, it is difficult to find influential factors that reflect the true sentiment of investors. Even for commodities like oil, China has a totally different price schedule, which is almost completely controlled by the government, regardless of the global price trend. As a result, we cannot make use of many common influential factors that work well in the developed countries. After some intensive search and testing, we finally select the following indicators that are suitable for the SSE.

1. *Shanghai Interbank Offered Rate (SHIBOR):*
   For a stock market, the interest rate is always the primary influential factor under consideration. The Shanghai Interbank Offered Rate aims to become a new interest rate benchmark in the China market, and to provide the similar functions and roles as those of the Federal Funds Rate (FFR) in the U.S. market and the London Inter Bank Offered Rate (LIBOR) in the U.K. market. Its maturity is important in making interest rates more market-based. Thus, researchers usually adopt SHIBOR to study the interaction between the interest rate and the China stock market. We use the changes between the daily SHIBOR overnight rate as the corresponding Interest Rate Indicator (IRI), which is defined similarly as that in (3.21):
$$\mathrm{SHIBOR}(n) = 100 \times \frac{i_c(n) - i_c(n-1)}{i_c(n-1)}, \tag{5.2}$$
   where i_c is the value of the SHIBOR overnight rate. Since China has just experimented the SHIBOR trial from October 2006, the IRI data are only available from then.

2. *International Stock Market Indicator (ISMI):*
   The interactions among stock markets around the world have become more and more intensified these days. The dynamic relationships between the China stock market and the stock markets in other countries have been extensively studied in the literature, among which the influence of the U.S. stock market is pervasive. In terms of daily stock returns, Laurencec *et al.* [82] found that the U.S. stock market has a strong causal effect to both the China and Hong Kong stock markets. In Chen *et al.* [28], it was reported that the Standard & Poor's 500 Index (S&P 500) led the SSE with respect to return transmission. Studies had also been conducted on other markets, which include the Hong Kong market [82, 142], the Japan market [134] and the India market [28]. In this study, we use the internal residues of the S&P 500 and Hong Kong Hang Seng Index (HSI), generated by our system adaptation framework, to represent the influences of the international stock markets on the SSE. More specifically, the internal residues of the S&P 500 and the HSI are generated by the system adaptation framework with the following respective internal OE models:

$$H_{\mathrm{S\&P500}}(z) = \begin{bmatrix} \frac{-45.94z^{-1} - 14.02z^{-2} + 9.758z^{-3} - 0.098z^{-4}}{1 + 0.2986z^{-1} - 0.2124z^{-2}} \\ \frac{12.62z^{-1} + 34.94z^{-2} - 37.14z^{-3} + 12.06z^{-4}}{1 - 0.8782z^{-1} + 0.2605z^{-2}} \\ \frac{2.489z^{-1} - 13.79z^{-2} - 3.118z^{-3} + 7.741z^{-4}}{1 + 0.1254z^{-1} - 0.5448z^{-2}} \end{bmatrix}^{\mathrm{T}}, \tag{5.3}$$

   and

$$H_{\mathrm{HSI}}(z) = \begin{bmatrix} \dfrac{2.418z^{-1}+2.45z^{-2}-0.108z^{-3}-0.6531z^{-4}}{1+1.158z^{-1}+0.1604z^{-2}} \\ \dfrac{0.0552z^{-1}-1.573z^{-2}+0.0189z^{-3}+1.519z^{-4}}{1+0.0662z^{-1}-0.9176z^{-2}} \\ \dfrac{2.573z^{-1}-1.364z^{-2}-3.253z^{-3}+1.912z^{-4}}{1+0.0694z^{-1}-0.9244z^{-2}} \end{bmatrix}^{\mathrm{T}} . \tag{5.4}$$

These OE models are respectively obtained by using their corresponding historical data for a period from 2001 to 2005.

3. *Exchange Rate of the U.S. Dollar against the Chinese Yuan (USD/CNY):*
China has reformed its currency policies in recent years, including a shift to a flexible exchange rate regime and pegging Chinese Yuan to a basket of foreign currencies rather than strictly tying to the U.S. dollar. Since then, the Chinese Yuan has been appreciated a lot against the U.S. dollar. It is allowed to float within a daily band of 0.5% around the central parity. This revaluation of CNY/USD marked the new era of managed floating exchange rate and had a lot of influences on the China stock market. Nieh and Yau [100] proved the existence of an asymmetric causal relationship between the appreciation of CNY/USD and the SSE. Similar causal relationships have also been confirmed by Yang [133] and Tian and Ma [126]. In our study, we select the change of the USD/CNY exchange rate as one of the influential factors for testing the Shanghai stock market.

4. *Inflation Rate Indicator (IFRI):*
The relationship between the inflation rate and the stock market is still debatable. Chow and Lawler [37] found that the higher mean rate of return in the SSE than that in the New York Stock Exchange Composite Index was partially the result of a higher rate of inflation in China. Huang *et al.* [69] employed multiresolution wavelet to investigate three influential factors in different timescales. In their studies, the inflation was found to have an impact on the SSE in the 16-to-32-month trend, but it vanished in the 2-to-4-month cycles. The inflation rate $\pi(n)$ is defined based on the logarithmic changes of Consumer Price Index (CPI) $C(n)$ (from the EIU country database), i.e.,

$$\pi(n) = 100 \times [\ln(C(n)) - \ln(C(n-1))] . \tag{5.5}$$

Since only the monthly CPI data are available, a cubic spline interpolation is used to increase its frequency. Inspired by the work of Huang *et al.* [69], we investigate the IFRI indicator at different frequencies, i.e., at the daily, weekly and monthly, respectively. It indeed turns out that the causal relationship changes at different frequencies.

5.2.2 Causality Tests

We first conduct a series of causality tests for the inflation rate indicator (IFRI). Specifically, we test the indicator respectively at the daily, weekly and monthly frequencies, and the results are shown in Figures 5.1 – 5.3. It is clear that the linear causal relationship is only significant at the weekly frequency. With the monthly data, although the inflation rate has significant causal effect to the SSE at the beginning, the causal relationship changes its direction after 2008. We further proceed to identify the nonlinear causal relationship using daily and monthly data. As shown in Table 5.1, there is no significant result found in the nonlinear tests. As such, the IFRI is not considered to be a final candidate for the input to the system adaptation framework of the SSE for daily prediction. Nonetheless, it could be used for weekly forecasting.

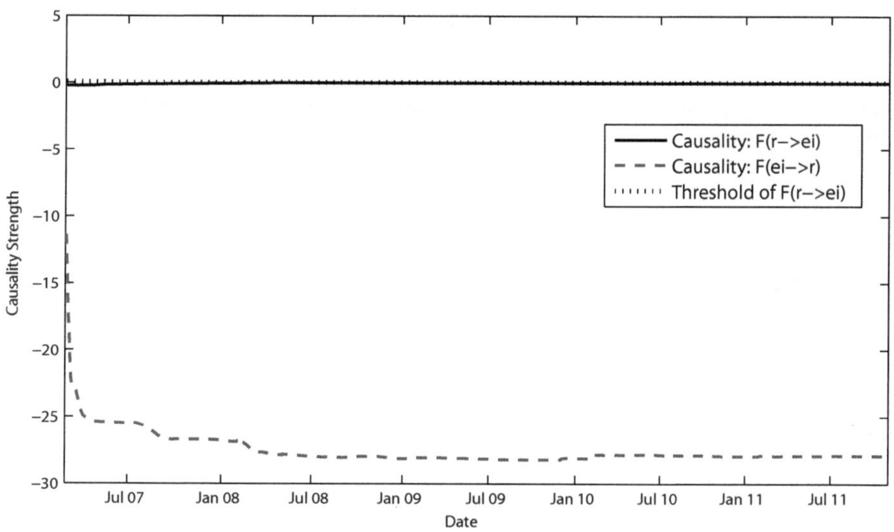

Fig. 5.1 Time-varying causality between the internal residue of the SSE and IFRI (daily data)

Data adjustment and data normalization are applied to the remaining four indicators before conducting causality tests. The lag length is selected in a similar way as it is in the DJIA (see Section 4.3). For the internal residue of the SSE, when the order reaches 4, the Durbin-Watson statistic is 2.00 and the p value of Breusch-Godfrey test is 0.44, both showing no autocorrelation in the residuals. As such, the lag lengths of the internal residues of the SSE, S&P 500 and HSI are all set to be 4 and for other influential factors, it is set to 10. Figures 5.4 – 5.7 present the time-varying causality between each of these indicators and the internal residue of the SSE. Only the S&P 500 after September 25, 2007, and the currency pair USD/CNY after January 30, 2008, significantly Granger cause the internal residue.

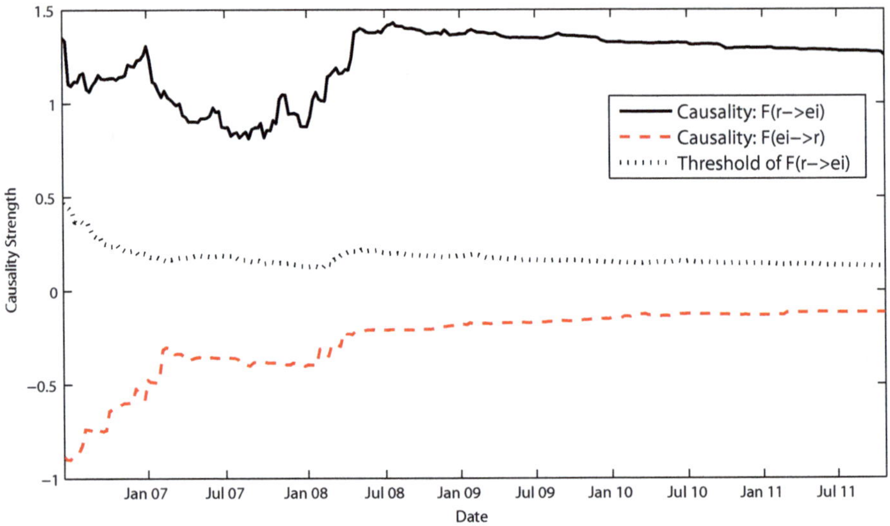

Fig. 5.2 Time-varying causality between the internal residue of the SSE and IFRI (weekly data)

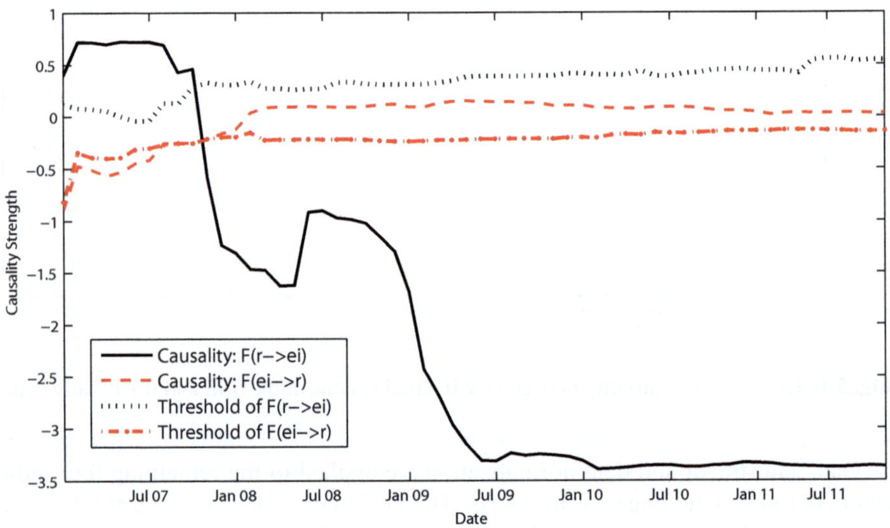

Fig. 5.3 Time-varying causality between the internal residue of the SSE and IFRI (monthly data)

Table 5.1 Nonlinear Granger causality test results between the internal residue of the SSE and IFRI at different frequencies

	Daily IFRI does not cause SSE internal residue		Monthly IFRI does not cause SSE internal residue	
Le = *Lr*	CS	TVAL	CS	TVAL
1	−0.0749	−1.4569	−0.0084	−0.6020
2	−0.1293	−2.0022	−0.0229	−1.5541
3	−0.2297	−2.9022	−0.0445	−1.6982
4	−0.2577	−2.9222	−0.0546	−1.0853
5	−0.2661	−3.7184	−0.1252	−1.8803
6	−0.3202	−3.6521	−0.2701	−2.3571
7	−0.5006	−2.9899	−0.1941	−1.1481
8	−0.3514	−1.7628	−0.1789	−0.6418
9	−0.6746	−40.6963	0.0765	1.1455
10	NA	NA	NA	NA

* Significance at 10% level for a one-sided test.
** Significance at 5% level for a one-sided test.

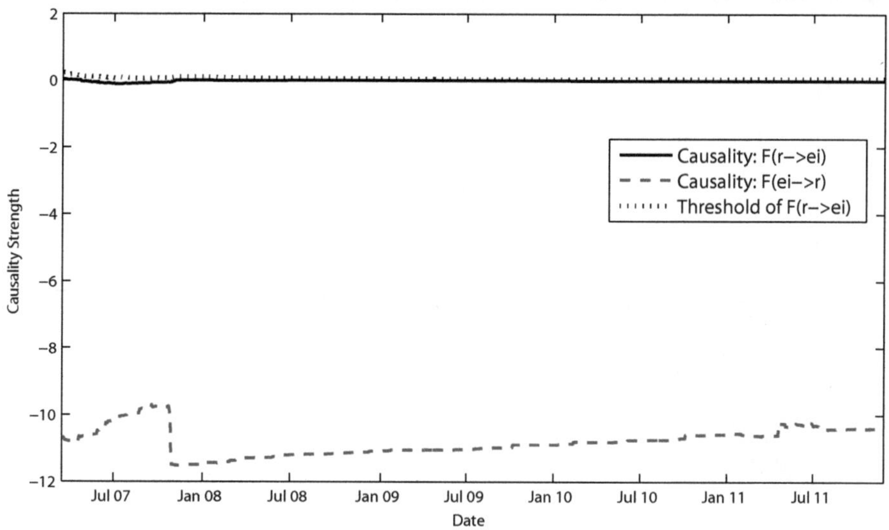

Fig. 5.4 Time-varying causality between the internal residue of the SSE and the IRI

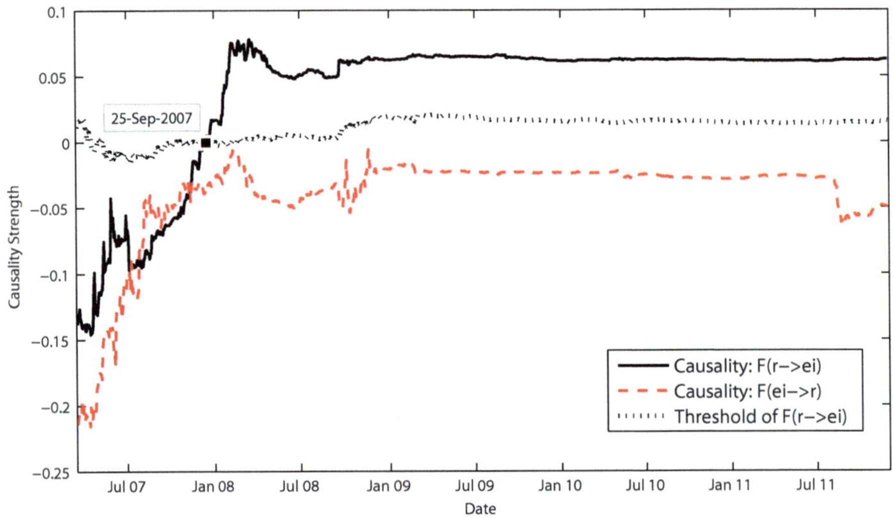

Fig. 5.5 Time-varying causality between the internal residues of the SSE and S&P 500

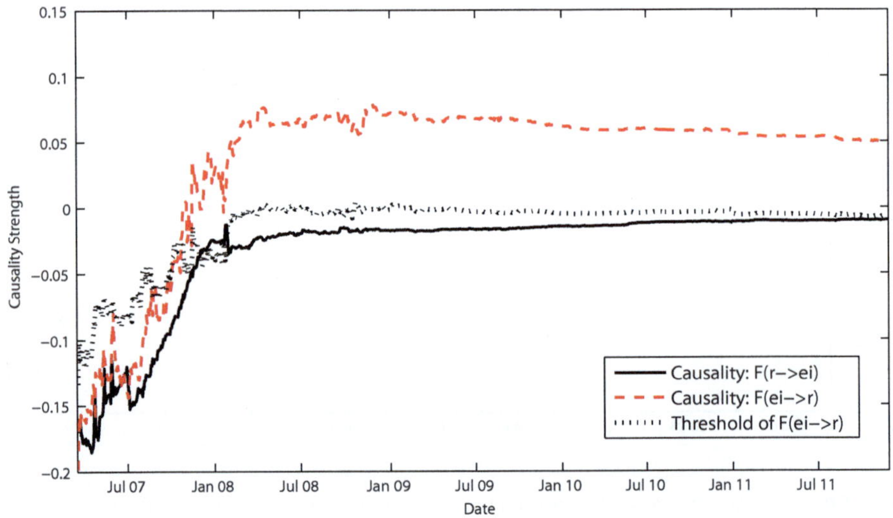

Fig. 5.6 Time-varying causality between the internal residues of the SSE and HSI

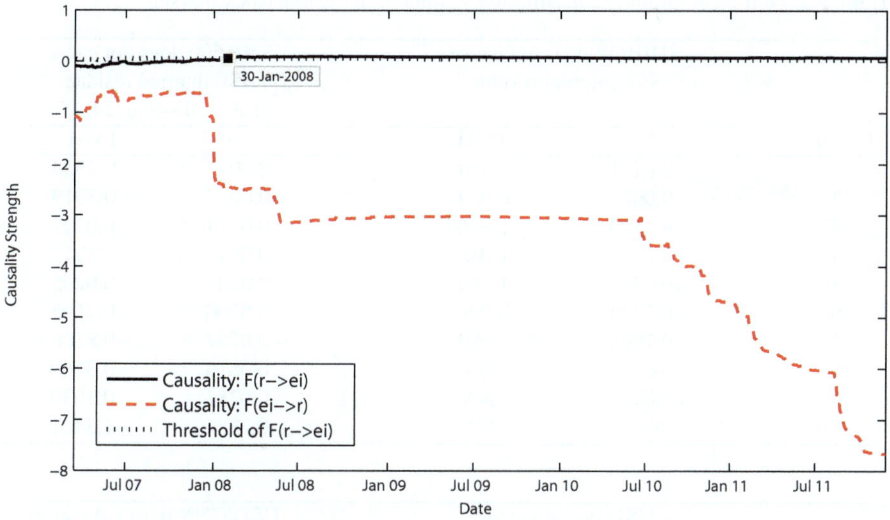

Fig. 5.7 Time-varying causality between the internal residues of the SSE and USD/CNY

We next apply the nonlinear causality test to the IRI and the HSI in the whole period; the S&P 500 from January 4, 2007, to September 24, 2007; and the USD/CNY from January 4, 2007, to January 29, 2008, and the results are given in Table 5.2, in which 'NA' denotes that results not available in the nonlinear causality tests. It is caused by $C_2 = 0$ or $C_4 = 0$ in (3.15) so that the condition in (3.18) cannot be tested. We note that $C_2 = 0$ or $C_4 = 0$ means that with selected lag length and given conditions, there are no vectors of e_{i} and r, whose distance is within ρ or less. It can be observed from the obtained results that significant nonlinear causality is only evidenced in the HSI with the lag length from 1 to 9. The nonlinear causal relationship is not found in the direction from the IRI, or the beginning parts of the S&P 500 and the USD/CNY, to the internal residue of the SSE.

Considering both the time-varying and nonlinear causality test results, we can conclude that the HSI over the whole testing period, the S&P 500 after September 25, 2007, and the USD/CNY after January 30, 2008, significantly Granger cause the internal residue of the SSE.

Lastly, we carry out the multicollinearity test for the S&P 500, the HSI and the USD/CNY after January 30, 2008. The resulting tolerances from January 30, 2008, to the end of November 2011, are given in Table 5.3. The tolerances are all close to 1, indicating little multicollinearity exists among them. As a result, all these three indicators are used as the final candidates for the input to the framework for the SSE from January 30, 2008, to the end of November 2011.

Table 5.2 Nonlinear Granger causality test results in the China stock market

	SHIBOR does not cause SSE internal residue		S&P 500 does not cause SSE internal residue (Jan. 4, 07–Sep. 24, 07)	
Le = *Lr*	CS	TVAL	CS	TVAL
1	0.0121	0.9849	−0.0017	−0.2824
2	0.0021	0.0572	−0.0037	−0.3319
3	−0.1590	−1.8365	0.0014	0.0721
4	−0.1892	−1.0867	−0.0042	−0.1917
5	−0.3318	−1.5631	0.0007	0.0205
6	−0.2217	−0.9902	−0.0360	−0.7195
7	−0.7098	−41.6938	−0.0747	−0.8306
8	NA	NA	−0.0816	−0.3760
9	NA	NA	−0.0287	−0.1039
10	NA	NA	−0.2381	−1.3421

	HSI does not cause SSE internal residue		USD/CNY does not cause SSE internal residue (Jan. 4, 07–Jan. 29, 08)	
Le = *Lr*	CS	TVAL	CS	TVAL
1	0.0091	3.0486**	−0.0024	−0.1740
2	0.0193	4.2702**	−0.0101	−0.4492
3	0.0261	4.1049**	−0.0204	−0.6314
4	0.0344	4.0747**	0.0130	0.2560
5	0.0329	3.2191**	0.0708	1.0750
6	0.0205	1.7307**	0.1196	0.9968
7	0.0274	2.0246**	0.5032	6.6323**
8	0.0221	1.3425*	0.5279	6.6009**
9	0.0255	1.3359*	NA	NA
10	0.0255	1.1054	NA	NA

* Significance at 10% level for a one-sided test.
** Significance at 5% level for a one-sided test.

Table 5.3 Multicollinearity test results in the China stock market

	S&P 500	HSI	USD/CNY
Tolerance	0.863	0.977	0.867

5.2.3 *Predicting Performance and Analysis*

As in the DJIA case, we partition the time interval of interest for the SSE into four subperiods according to different phases of the market and the variance of its internal residue. The training period for estimating the adaptive filter is from January 30, 2008, to the end of April 2008. Subperiod S1 is from May to December 2008, a period characterized as a steep decline. From January 2009, the China stock market began to recover until July 2009. Then, it came into an oscillation period. As such, Subperiod S2 is set to be from January to July 2009. The oscillation period, from August 2009 to December 2010, is Subperiod S3. In 2011, the market turned down again. We therefore define Subperiod S4 as from January 2011 to November 2011. Figures 5.8 and 5.9, respectively, display the subperiods in the SSE daily closing prices and its internal residue.

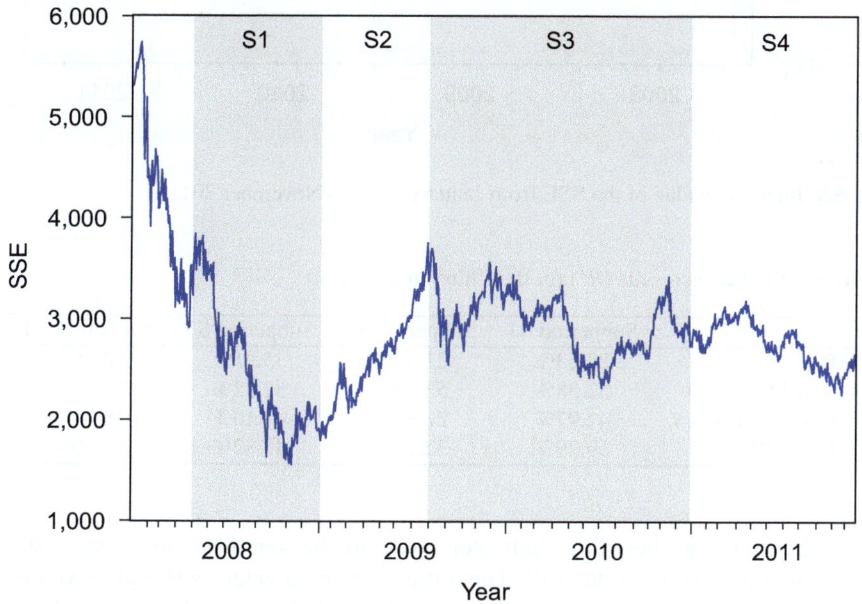

Fig. 5.8 Daily closing prices of the SSE from January 2008 to November 2011

The adjusted R^2 (i.e., $\bar{R}^2$) values for the SSE are given in Table 5.4. The initial hyperparameters in the test of $\bar{R}^2$ for each indicator and in the final prediction with the complete input set are the same. This rule is the same as those adopted for the U.S. market. More specifically, the hyperparameters of the AR part are fixed and their initial values are set as 5.5×10^{-4}. Initial covariance matrix $\hat{P}$ is set as a diagonal matrix with all its diagonal entries equal to 10^5. It is clear that the S&P 500 is the most dominant factor in all the subperiods, implying that the China stock market has higher potential to be influenced by the U.S. market.

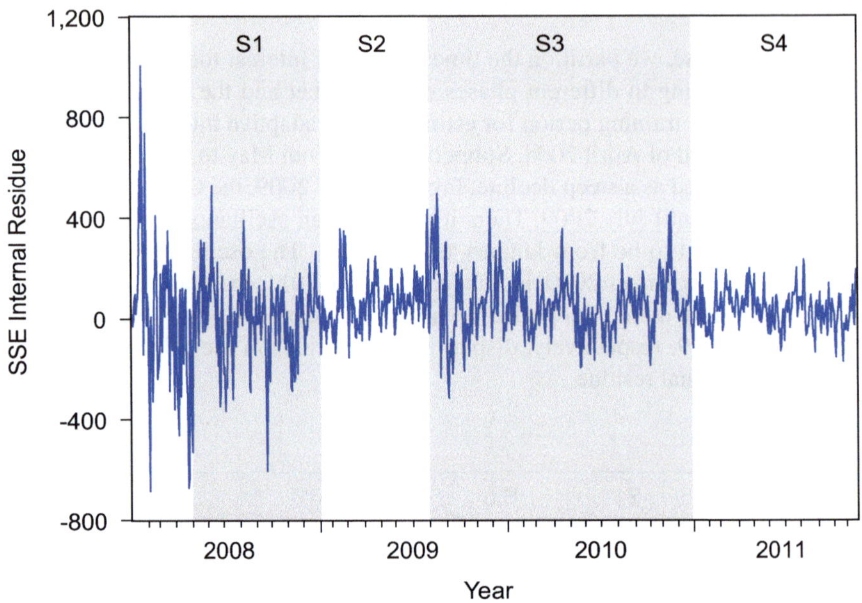

Fig. 5.9 Internal residue of the SSE from January 2008 to November 2011

Table 5.4 Prediction results ($\bar{R}^2$) for the China stock market

	Subperiod S1	Subperiod S2	Subperiod S3	Subperiod S4
AR	22.63%	21.77%	35.91%	22.68%
AR and S&P 500	52.38%	53.35%	52.22%	53.31%
AR and USD/CNY	42.97%	22.87%	43.10%	38.45%
AR and HSI	39.29%	32.25%	45.82%	37.40%

We then combine these three indicators to obtain the adaptive filter for the system adaptation framework for the SSE. The initial hyperparameters in the adaptive filter (the diagonal entries of Q_r) are also set to be 5.5×10^{-4} and other initial conditions are set to be the same as those for the DJIA. The resulting estimations of the hyperparameters in the adaptive filter, i.e., the coefficients associated with autoregressive part, are

$$4.6958\times10^{-5},\ 3.1908\times10^{-5},\ 6.2916\times10^{-7},\ 7.7498\times10^{-6}, \tag{5.6}$$

the estimated coefficients associated with the S&P 500 input are

$$5.1311\times10^{-4},\ 0.0980,\ 0.2599,\ 0.3537, \tag{5.7}$$

the estimated coefficients associated with the HSI input are

$$0.1398,\ 4.2367\times10^{-4},\ 3.7451\times10^{-4},\ 0.1066, \tag{5.8}$$

and finally, the estimated coefficients associated with the USD/CNY input are

$$8.1674\times10^{-5},\ 0.2927,\ 5.1603\times10^{-7},\ 7.5472\times10^{-4},\ 7.9529\times10^{-5}, \tag{5.9}$$

$$0.0114,\ 6.6762\times10^{-6},\ 8.6714\times10^{-5},\ 6.5181\times10^{-5},\ 6.2095\times10^{-8}. \tag{5.10}$$

With the estimated hyperparameters, the one-day-ahead prediction results are obtained and shown in Table 5.5. It is once again confirmed that our framework gives more accurate one-step-ahead prediction than the conventional ARMAX approach. For illustration, we calculate the average daily changes of the SSE closing prices in the four subperiods, which are 53.60, 35.63, 36.26 and 23.30, respectively. The MAE of the ARMAX model is about the same as the average daily changes, whereas the MAE resulting from our framework is much smaller and much reasonable. Nevertheless, we should also note that the prediction results for the China market are not as good as those for the U.S. market. One possible reason could be due to the inaccurate data that we can gather from the open sources. Another possible way to enhance the prediction performance is to incorporate the influence of the Chinese Government, if possible, as an input to the system adaptation framework. The China stock market is clearly interfered by some factors beyond the common market influential sources. For example, the interest rate is believed to be the most important and direct factor that affects a stock market, but the SHIBOR, a market-based interest rate that reflects the tightness of market liquidity, does not Granger cause the internal residue of the SSE at all. Last but not the least, the China stock market is still not well regulated and is too immature to have sentiment indicators as which have seriously affected the developed markets. This is also a common feature of emerging markets.

5.3 Hong Kong Hang Seng Index

As one of the major international financial centers, Hong Kong is a well-developed market with free economy as well as effective and transparent regulations. The Hang Seng Index (HSI) is a market capitalization-weighted index of the Hong Kong Stock Exchange, which covers about 65% of its total capitalization. It is one of the most widely quoted indices in Asia and it behaves as a leading indicator of the performance of the general economy in Hong Kong. Many investors use the HSI as their performance benchmark while researchers usually consider it as a representation of the stock markets in Asia. The Hong Kong stock market is one of the most active and fastest growing markets in the world, in which a wide variety of products are offered and traded, providing many useful influential factors and market data. In our study, we are going to take a close examination of the closing prices of the HSI from the beginning of 2006 to November 2011. The internal model for the HSI was obtained earlier in (5.4), which was identified using the HSI data from 2001 to 2005, and was

Table 5.5 Comparison of the prediction results between the ARMAX approach and the proposed framework for the China stock market

	Subperiod S1 May 2008–Dec. 2008		Subperiod S2 Jan. 2009–Jul. 2009	
	MAE	RMSE	MAE	RMSE
ARMAX	52.00	67.47	34.44	42.12
System adaptation framework	19.63	30.30	16.96	24.56
Improvement (%)	62.26	55.09	50.75	46.75

	Subperiod S3 Aug. 2009–Dec. 2010		Subperiod S4 Jan. 2011–Nov. 2011	
	MAE	RMSE	MAE	RMSE
ARMAX	36.30	49.60	23.70	30.41
System adaptation framework	26.56	38.13	12.79	18.22
Improvement (%)	26.82	23.13	46.05	40.07

used to generate the internal residue of the HSI, an influential factor selected for the China stock market in the previous section.

5.3.1 Input Selection

High degree of liquidity makes the Hong Kong stock market very sensitive to the external factors [129]. Garefalakis *et al.* [50] explored the effects of various economic indicators on the HSI, among which the S&P 500, crude oil, volatility of the gold returns and USD/JPY are four determinant factors. Yu and Tam [138] monitored the investors' sentiment of the Hong Kong stock market by the so-called Hang Seng China Enterprise Index. In our study, we consider the following economic and sentiment factors: the interest rate, the international stock markets, the foreign exchange rate, the oil price and the sentiment index.

1. *HKD Interest Settlement Rates (HKDISR):*
   The interest rate is always considered as a primary influential factor in a well-developed stock market. For the Hong Kong market, we adopt the changes of the HKD interest settlement rate as the interest rate indicator. As part of the Hong Kong Association of Banks (HKAB) Forward Rate Agreement terms, the daily HKD interest settlement rate is fixed by reference to the market rate for HKD deposits in the Hong Kong interbank market[1]. The HKAB provides quotations from 20 banks as reference to calculate the HKDISR. Daily fixings are made available for various HKD deposit maturity, in which we particularly choose overnight maturity for the HKDISR. The daily change of the HKDISR, i.e.,

[1] The definition is from the Hong Kong Association of Banks. More information can be found at http://www.hkab.org.hk/.

$$\mathrm{HKDISR}(n) = 100 \times \frac{i_k(n) - i_k(n-1)}{i_k(n-1)}, \tag{5.11}$$

is used as an influential indicator for the HSI, in which i_k is the overnight HKD interest settlement rate.

2. *International Stock Market Indicator (ISMI):*
The Hong Kong stock market has great tendency to be affected by other international stock markets, especially those from the U.S. and China. The studies about the comovement of the international stock markets could be frequently found in the literature. Among them, the U.S. stock market has been found to exert its influence to the markets worldwide [38, 113]. Its influence on the China stock market has also been testified earlier in Section 5.2, in which it has been shown that the HSI also presents its significant causality effect to the SSE. With the rapid increase of economic integration between Hong Kong and China, the dynamic relationship between the Hong Kong and China stock markets has attracted more and more attention. In our studies, we use the S&P 500 and the SSE to represent the influences of the international stock markets over the HSI. As usual, the internal residue of the S&P 500 is obtained by the system adaptation framework with the OE model given in (5.3). Similarly, the daily closing prices of the SSE from 2001 to 2005 are used to estimate its corresponding internal OE model, which is given as

$$H_{\mathrm{SSE},0105}(z) = \begin{bmatrix} \dfrac{10.7z^{-1}+3.068z^{-2}-4.686z^{-3}-1.002z^{-4}}{1-0.6288z^{-1}+0.1369z^{-2}} \\ \dfrac{-3.491z^{-1}-3.564z^{-2}-5.286z^{-3}+6.115z^{-4}}{1-0.6835z^{-1}+0.01461z^{-2}} \\ \dfrac{-0.0132z^{-1}+0.8774z^{-2}-1.472z^{-3}+0.6056z^{-4}}{1-1.468z^{-1}+0.5217z^{-2}} \end{bmatrix}^{\mathrm{T}}. \tag{5.12}$$

The internal residue of the SSE under the system adaptation framework can then be generated accordingly.

3. *Exchange Rate of the Hong Kong Dollar against the U.S. Dollar (HKD/USD):*
The currency exchange rate has shown its connection with the stock markets in the U.S. and China. Hong Kong has implemented a linked exchange rate system between the Hong Kong dollar (HKD) and the U.S. dollar (USD) since 1983. Under this system, the HKD was pegged to the USD but allowed to be traded within a certain range. The link between the exchange rate and stock price in Hong Kong has been examined by Yu [139], Phylaktis and Ravazzolo [105] and many others. We adopt in our studies the daily change of the HKD/USD pair to measure the influence of the exchange rate to the Hong Kong market. The daily change is calculated in the same fashion as that in (5.11).

4. *Oil Price (OP):*
   Crude oil is an important commodity, which is heavily traded by both businesses and speculators. As it has been proven in Chapter 4, the oil price also has influences on the U.S. stock market. However, there is no consensus about such a relationship in other places, especially in the Great China region. Nonetheless, we select the oil price as an influential indicator for the Hong Kong market, as that used to investigate the U.S. market in Section 3.3.

5. *Baltic Dry Index (BDI):*
   Hong Kong is a major port in the world, and is also the logistics gateway to China. It is one of the key factors contributing to the economic growth of Hong Kong. As such, we consider the Baltic Dry Index (BDI), the same as that in Section 3.3, as the indicator to measure the effect that the global shipping price has on the Hong Kong stock market.

6. *HSI Volatility Index (VHSI):*
   In January 2011, the HSI Volatility Index (VHSI) was issued by Hang Seng Indexes Company Limited, aiming to measure market expectations of the volatility in the HSI over the next 30-day period. Acting similarly as the VXD to the DJIA, the VHSI is considered to be an important sentiment indicator to the Hong Kong stock market. We use its change, calculated similarly as in (5.11), as an influential factor in the framework related to the HSI.

5.3.2 Causality Tests

All the seven selected indicators are first preprocessed by data adjustment and data normalization processes before conducting necessary causality tests. As in the earlier tests, the lag length is set to be 4 for the internal residues of the S&P 500 and the SSE, and 10 for the other influential factors. The resulting time-varying causality relationships between the indicators and the internal residue of the HSI are presented in Figures 5.10 – 5.16. Among the seven indicators, only the S&P 500 significantly Granger causes the internal residue of the HSI during the whole period of interest, i.e., January 2006 to November 2011.

We then further proceed to carry out nonlinear causality tests to those indicators in the periods when no significant linear causality is observed. More specifically, the nonlinear causality tests are conducted for the HKDISR, the SSE and the HKD/USD in the whole period, for the OP from January 24, 2006 to June 9, 2008, for the BDI from January 24, 2006 to February 4, 2008, and lastly, for the VHSI from January 24, 2006 to May 21, 2006. Table 5.6 gives all the nonlinear causality testing results. We note that significant nonlinear causal effects from the HKDISR and the SSE are found with all the lag lengths, whereas no noticeable effect is observed for the other indicators.

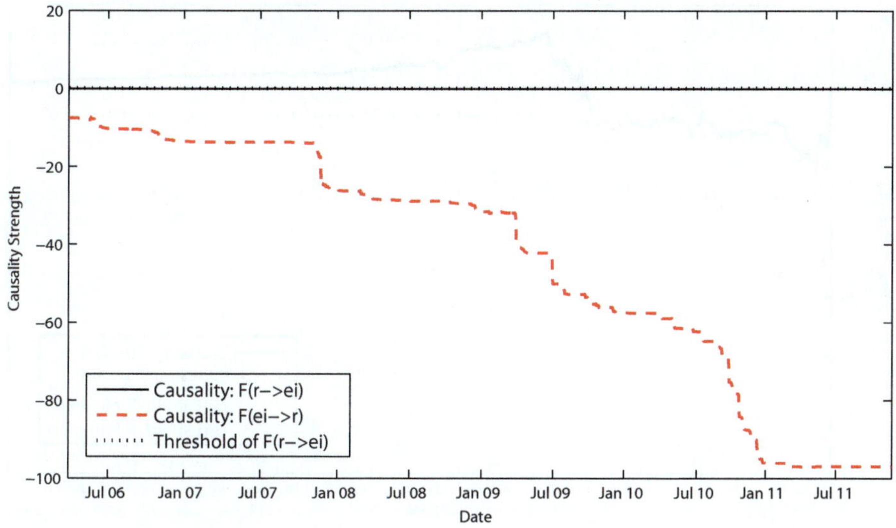

Fig. 5.10 Time-varying causality between the internal residue of the HSI and the HKDISR

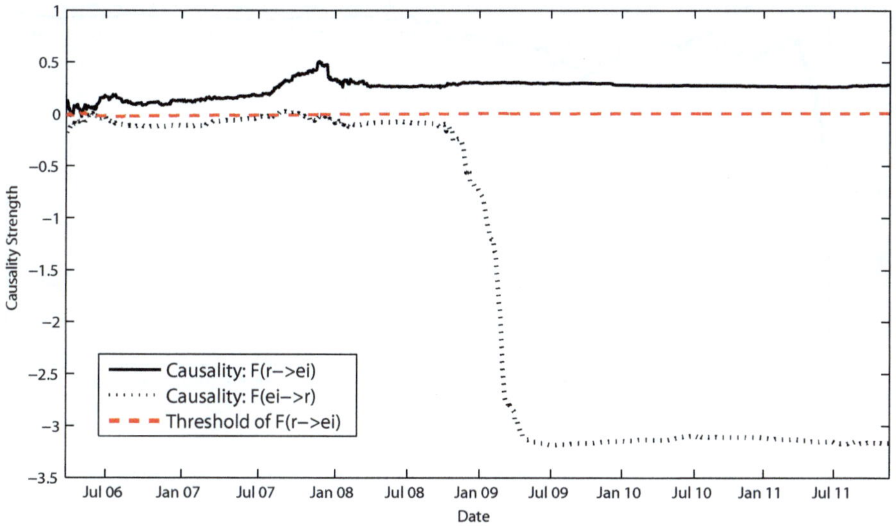

Fig. 5.11 Time-varying causality between the internal residues of the HSI and S&P 500

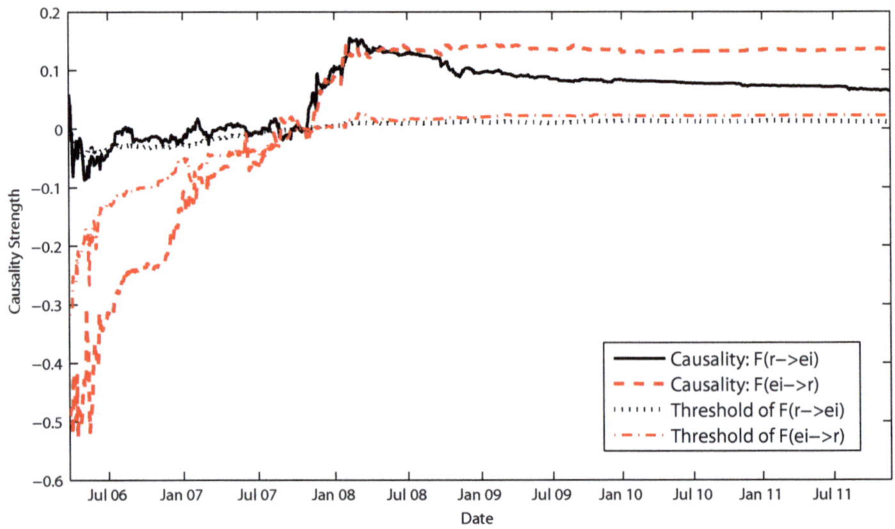

Fig. 5.12 Time-varying causality between the internal residues of the HSI and SSE

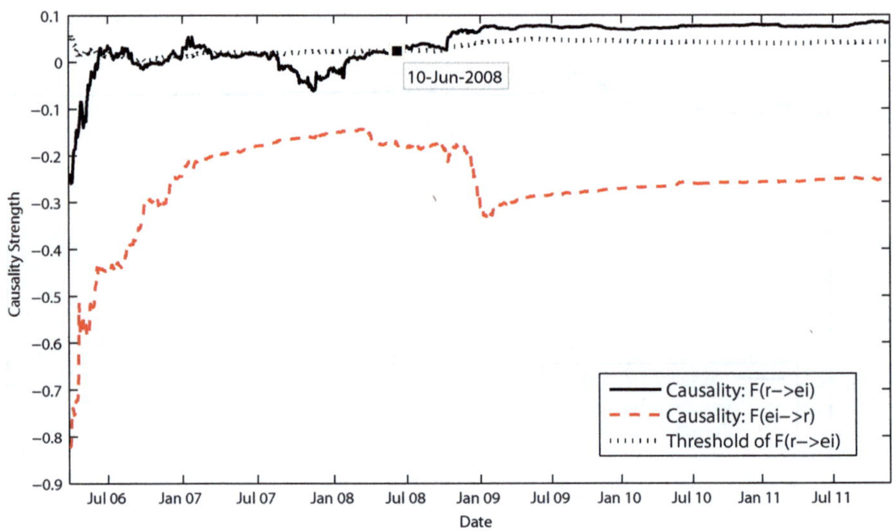

Fig. 5.13 Time-varying causality between the internal residue of the HSI and the OP

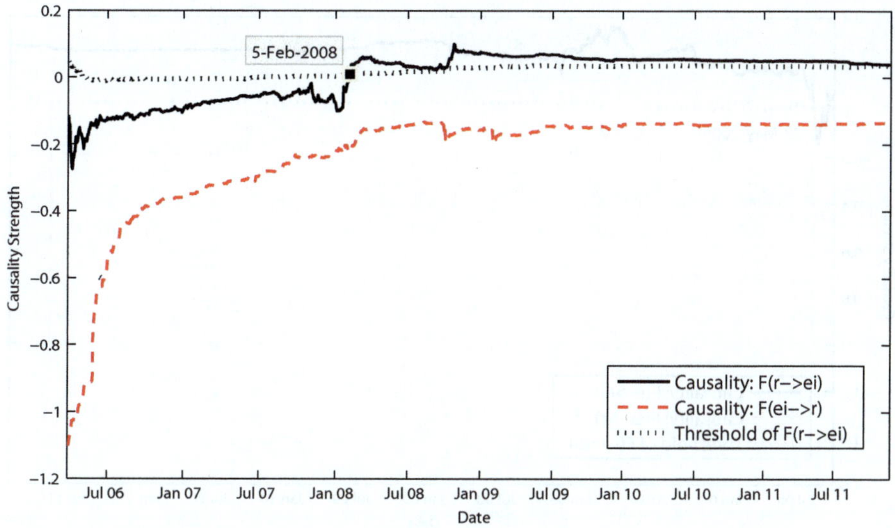

Fig. 5.14 Time-varying causality between the internal residue of the HSI and the BDI

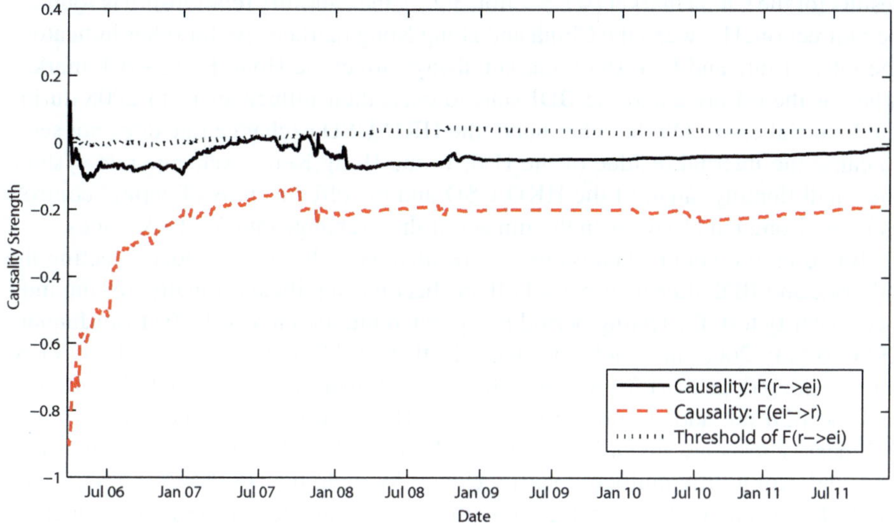

Fig. 5.15 Time-varying causality between the internal residue of the HSI and the HKD/USD

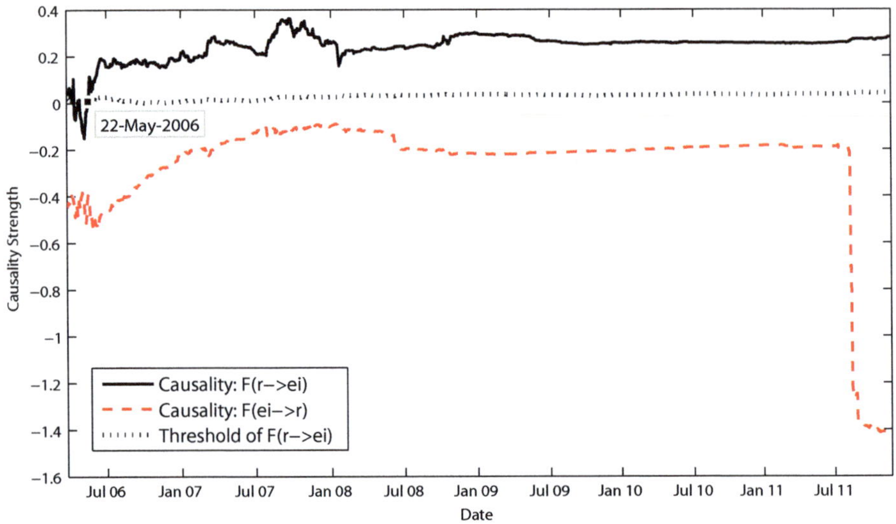

Fig. 5.16 Time-varying causality between the internal residue of the HSI and the VHSI

According to our tests, it is found that the Hong Kong stock market is sensitive to various influential factors especially during the crisis period. Compared with the test results for the China market (see Section 5.2.2), the causality relationship is found to be bidirectional between the China and Hong Kong markets. As for other indicators, the interest rate and U.S. stock market always affect the Hong Kong stock market, whereas the oil price and the BDI start to exert their influences from 2008 during the financial crisis. On the other hand, the HKD/USD exchange rate does not seem to cause the internal residue of the HSI. In the Hong Kong exchange rate system, the small floating range of the HKD/USD and the effectiveness of capital controls in Hong Kong tend to weaken the impact of the exchange rate on stock prices.

Based on the obtained causality test results, we will only consider selecting the oil price and BDI after their causal effects become significant. Finally, for the multicollinearity test, the testing period is separated into two intervals: (i) from January 2006 to May 2008, in which only the HKDISR, S&P 500, the SSE and the VHSI are considered; and (ii) from June 2008 to November 2011, for which the oil price and the BDI are added into the regression. The resulting tolerances of the multicollinearity tests are given in Table 5.7. We note that all the tolerances are close to 1, indicating little multicollinearity exists. As such, the HKDISR, the S&P 500, the SSE and the VHSI are selected as the final candidates of the input influential factors for the system adaptation framework of the HSI for the whole period of interest, whereas the OP and the BDI are added in for the interval from June 2008 to November 2011.

Table 5.6 Nonlinear Granger causality test results in the Hong Kong stock market

	HKDISR does not cause HSI internal residue		SSE does not cause HSI internal residue		HKD/USD does not cause HSI internal residue	
Le = *Lr*	CS	TVAL	CS	TVAL	CS	TVAL
1	0.1860	6.5155**	0.0035	2.5165**	0.0044	1.4349**
2	0.2601	15.5890**	0.0076	3.4308**	0.0117	1.9236**
3	0.2757	16.1359**	0.0136	3.7639**	0.0101	1.1906
4	0.2771	16.3122**	0.0154	3.9318**	4.77×10^{-4}	0.0453
5	0.2658	8.3784**	0.0167	3.6297**	−0.0103	−0.7564
6	0.2728	8.7427**	0.0202	3.6560**	−0.0111	−0.6131
7	0.2510	4.6947**	0.0296	4.7418**	−0.0051	−0.2347
8	0.2614	4.8131**	0.0315	4.8754**	−0.0183	−0.6808
9	0.2292	3.2622**	0.0377	5.2393**	−0.0215	−0.6899
10	0.2988	12.0728**	0.0363	5.0697**	−0.0188	−0.4516

	OP does not cause HSI internal residue (Jan. 24, 06–Jun. 9, 08)		BDI does not cause HSI internal residue (Jan. 24, 06–Feb. 4, 08)		VHSI does not cause HSI internal residue (Jan. 24, 06–May 21, 06)	
Le = *Lr*	CS	TVAL	CS	TVAL	CS	TVAL
1	0.0036	1.0569	0.0020	1.0748	-3.80×10^{-5}	−0.7029
2	0.0056	0.9218	-1.85×10^{-4}	−0.1473	-4.68×10^{-5}	−0.0944
3	0.0070	0.8613	−0.0014	−0.7709	0.0010	0.3497
4	0.0053	0.5528	−0.0023	−1.0773	0.0058	1.2675
5	0.0011	0.0936	−0.0032	−1.2998	0.0113	1.6302*
6	−0.0132	−0.9284	−0.0037	−1.2820	0.0200	2.0426**
7	−0.0264	−1.5057	−0.0042	−1.2940	0.0291	2.1242**
8	−0.0481	−2.1697	−0.0052	−1.2619	0.0137	0.7681
9	−0.0599	−1.5526	−0.0067	−0.0067	0.0291	0.9709
10	−0.1272	−2.2279	−0.0050	−0.9088	0.0394	1.1321

* Significance at 10% level for a one-sided test.

** Significance at 5% level for a one-sided test.

Table 5.7 Multicollinearity test results in the Hong Kong stock market

	Tolerance	
	Jan. 2006–May 2008	Jun. 2008–Nov. 2011
HKDISR	0.992	0.990
S&P 500	0.907	0.782
SSE	0.967	0.969
OP	–	0.906
BDI	–	0.980
VHSI	0.913	0.816

5.3.3 Predicting Performance and Analysis

For the HSI, we divide the whole period under investigation into five subperiods. The partition of subperiods is illustrated in the daily closing prices of the HSI depicted in Figure 5.17 and the resulting internal residue given in Figure 5.18. It can be observed from Figure 5.18 that the variance of the internal residue began to increase from July 2007, which is corresponding to a sharp rise of the HSI and is the time when the global financial crisis began to emerge. Affected by the crisis, the HSI started to crash in November 2007. Our prediction begins from July 2007. Since the BDI and the oil price began to affect the HSI after February 2008 and June 2008, respectively, the first subperiod (S1) is thus set from July 2007 to January 2008 (the training data are set from July 2006 to June 2007 for the initial period). In order to obtain meaningful results, we set training data sizes to be no less than half a year. As a result, the second subperiod (S2) is chosen to be from February 2008 to July 2008. The HSI entered into its rally since the beginning of 2009 and the variance of the internal residue became much smaller. The third subperiod (S3) is therefore set from August to the end of 2008. As it will be reported later in Chapter 6, there is a market turning period detected on around November 2010 for the Hong Kong market, we hence set the fourth subperiod (S4) to be from January 2009 to the end of 2010, and lastly, the fifth subperiod (S5) is from January 2011 to November 2011.

We should note that the partition of the market subperiods is subjective. There are no unique rules and there is no unique partition. Roughly speaking, it represents changes in either market dynamics and/or influential factors. Generally, in real-time applications and prediction, one should consider changing the period partition, equivalently changing either the adaptive filter and/or the OE model when observing a divergent trend in the predicting result. Once again, it is an interesting direction worth further investigation.

The training data used for obtaining the corresponding adaptive filter for the five subperiods are illustrated in Table 5.8. Initially, a set of one-year data prior to Subperiod S1 is used for training purpose. After Subperiod S1, the HSI ended its mildly bearish trend and then entered into a drastic fluctuation stage with a rapid rise followed by a crash. In Chapter 6, both market turning period and the structural breaks of macroeconomic indicators are detected in Subperiod S1. Therefore, although the influential indicators for Subperiods S1 and S2 are the same, we update the training data set for Subperiod S2 to include those data in Subperiod S1, i.e., the data from January 2007 to January 2008. The BDI and the OP data are added to estimate the adaptive filter in Subperiods S3 and S4. The initial hyperparameters in the adaptive filter (the diagonal entries of Q_r) for different subperiods are also given in Table 5.8. Other initial conditions are the same as those in the U.S. and China markets.

The MAE and RMSE are again used to measure the predicting performances. We first compare the predicting performances in Subperiods S1 and S2 by adopting different training periods, in which the initial hyperparameters are set to be the same. The results given in Table 5.9 show that the MAE of Subperiod S2 is remarkably reduced by using the new training period. Including data in the drastically fluctuated period shows its effect in predictions.

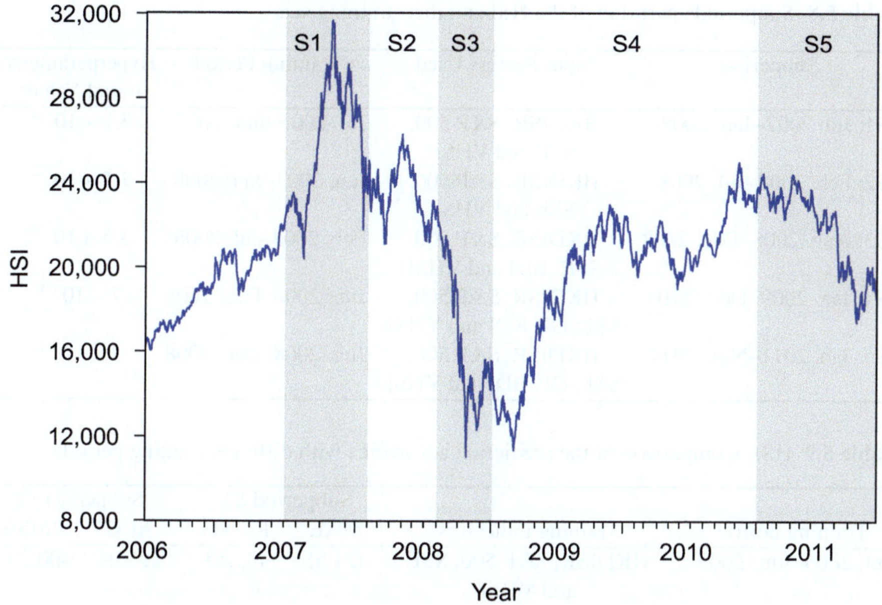

Fig. 5.17 Daily closing prices of the HSI from July 2006 to November 2011

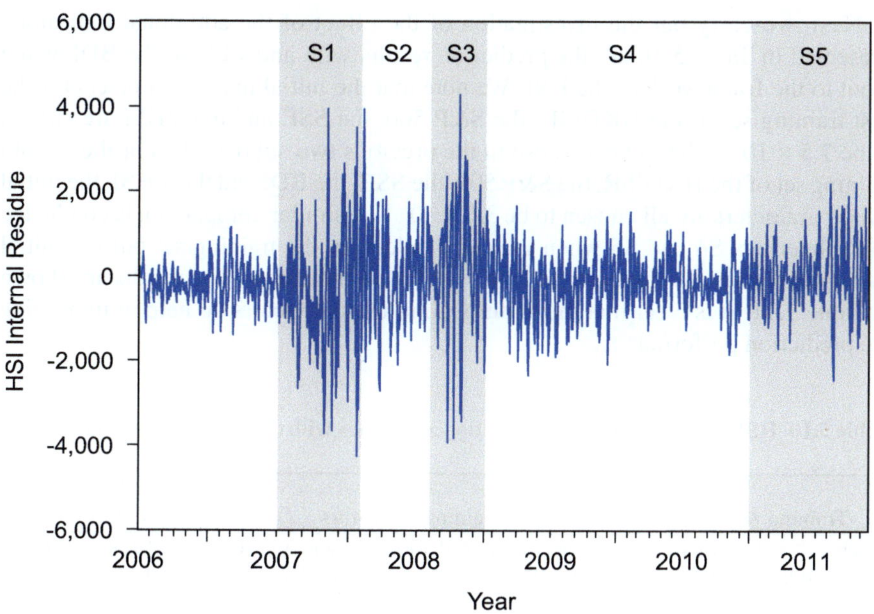

Fig. 5.18 Internal residue of the HSI from July 2006 to November 2011

Table 5.8 Subperiods partition of the HSI and their training sets

Subperiod	Input Factors Used	Training Period	Hyperparameter Initial Values
S1: Jul. 2007–Jan. 2008	HKDISR, S&P 500, SSE and VHSI	Jul. 2006–Jun. 2007	7.5×10^{-4}
S2: Feb. 2008–Jul. 2008	HKDISR, S&P 500, SSE and VHSI	Jan. 2007–Jan. 2008	7.5×10^{-4}
S3: Aug. 2008–Dec. 2008	HKDISR, S&P 500, SSE, BDI and VHSI	Feb. 2008–Jul. 2008	3.5×10^{-4}
S4: Jan. 2009–Dec. 2010	HKDISR, S&P 500, SSE, OP, BDI and VHSI	Jun. 2008–Dec. 2008	7×10^{-4}
S5: Jan. 2011–Nov. 2011	HKDISR, S&P 500, SSE, OP, BDI and VHSI	Jun. 2008–Dec. 2008	7×10^{-4}

Table 5.9 HSI: Comparison of the prediction accuracies with different training periods

		Subperiod S1		Subperiod S2	
Training period	Training indicators	MAE	RMSE	MAE	RMSE
Jul. 2006–Jun. 2007	HKDISR, S&P 500, SSE and VHSI	324.58	433.55	296.94	400.28
Jan. 2007–Jan. 2008	HKDISR, S&P 500, SSE and VHSI			108.18	152.72

Next, we carry out the investigation of the effect of the additional indicators. Presented in Table 5.10 are the prediction results with and without the BDI as the input to the framework of the HSI. We note that the initial hyperparameters for the first training set of the HKDISR, the S&P 500, the SSE and the VHSI are chosen to be 7.5×10^{-4}, the same as those in the previous two subperiods. For the second training set of the HKDISR, the S&P 500, the SSE, the BDI and the VHSI, the initial hyperparameters are all chosen to be 3.5×10^{-4}. A similar comparison is conducted for Subperiods S4 and S5 when the OP is added to the training set but the initial hyperparameters are chosen to be 7×10^{-4} for all indicators. The results are shown in Table 5.11. It is clear that the additional influential factors do help in improving the prediction performance.

Table 5.10 HSI: Comparison of the prediction accuracies with different training indicator set

		Subperiod S3	
Training period	Training indicators	MAE	RMSE
Feb. 2008–Jul. 2008	HKDISR, S&P 500, SSE and VHSI	219.56	296.58
Feb. 2008–Jul. 2008	HKDISR, S&P 500, SSE, VHSI and BDI	144.18	185.28

Table 5.11 HSI: Comparison of the prediction accuracies with different training indicator set

Training period	Training indicators	Subperiod S4 MAE	Subperiod S4 RMSE	Subperiod S5 MAE	Subperiod S5 RMSE
Jun. 2008–Dec. 2008	HKDISR, S&P 500, SSE and VHSI	222.08	301.54	199.63	261.36
Jun. 2008–Dec. 2008	HKDISR, S&P 500, SSE, VHSI and BDI	131.11	184.08	153.38	203.20
Jun. 2008–Dec. 2008	HKDISR, S&P 500, SSE, VHSI, BDI and OP	93.95	128.70	105.09	143.72

Finally, the ARMAX model is once again employed for benchmarking the predicting performance. The lag lengths of the ARMAX model are the same as those used in the U.S. and China market, i.e., 4 for the AR and MA terms and 10 for all the exogenous inputs. The prediction results provided by the ARMAX are under the same influential indicator set and same training period as those used in our framework. The one-step-ahead predicting performances are summarized in Table 5.12, which show that our framework again gives better results. The largest improvements are seen in Subperiods S2 and S3 when the market crashed. The result is consistent with that in other markets. It is again confirmed that the proposed system adaptation framework can effectively adapt to capture drastic changes in the stock markets. For easy references, we note that the average daily changes of closing prices in these five subperiods are 561.94, 367.32, 511.42, 248.47 and 249.00, respectively.

Table 5.12 Comparison of the prediction accuracies between the ARMAX approach and the proposed framework for the Hong Kong stock market

	Subperiod S1		Subperiod S2		Subperiod S3	
	MAE	RMSE	MAE	RMSE	MAE	RMSE
ARMAX	515.64	642.93	350.71	452.52	472.71	611.57
System adaptation framework	324.58	433.55	108.18	152.72	144.18	185.28
Improvement (%)	37.05	32.57	69.15	66.25	69.50	69.70

	Subperiod S4		Subperiod S5	
	MAE	RMSE	MAE	RMSE
ARMAX	238.45	306.37	234.93	303.98
System adaptation framework	93.95	128.70	105.09	143.72
Improvement (%)	60.73	58.00	55.27	52.72

5.4 Singapore Straits Times Index

Singapore and Hong Kong have been competing for the second largest international financial center in the Asia-Pacific region for decades. The Singapore stock market is lagging behind Hong Kong's in liquidity and market value of companies listed. However, in terms of listing international companies, Singapore is better. In this section, we are going to investigate the influential factors of the Singapore Straits Times Index (STI), which is a capitalization-weighted stock market index of 30 companies listed in the Singapore Stock Exchange (SGX). We will examine the behavior of the daily closing prices of the STI from the beginning of 2006 to November 2011. The corresponding internal residue of the STI is obtained through the system adaptation framework with the following OE model:

$$H_{\mathrm{STI}}(z)=\begin{bmatrix} \dfrac{1.573z^{-1}-1.158z^{-2}+0.3141z^{-3}-0.0497z^{-4}}{1-0.7621z^{-1}+0.5226z^{-2}} \\ \dfrac{1.056z^{-1}-0.316z^{-2}+0.3522z^{-3}-0.3381z^{-4}}{1-0.4146z^{-1}+0.076z^{-2}} \\ \dfrac{-0.0034z^{-1}-0.2253z^{-2}+0.1168z^{-3}-0.2021z^{-4}}{1-1.081z^{-1}+0.3826z^{-2}} \end{bmatrix}^{\mathrm{T}}, \tag{5.13}$$

which is estimated using the daily closing prices of the STI from 2001 to 2005.

5.4.1 Input Selection

The factors that drive the stock market of Singapore have also attracted many attentions during the last two decades. Mookerjee and Yu [98] studied the linkage between four macroeconomic variables (i.e., the narrow money supply, broad money supply, exchange rate and foreign exchange reserves) and the Singapore stock market based on the techniques of cointegration, causality and forecasting equations. A similar study incorporating more economic variables was conducted by Maysami and Koh [95], in which significant effects from changes in interest rates and exchange rates as well as the U.S. and Japanese stock markets are testified. In our study, we explore the dynamic relations between the STI and its potential influential factors including the interest rate, three international stock markets, the exchange rate, the oil price and the BDI.

1. *Singapore Overnight Rate Average (SORA):*
   The interest rate of Singapore moves closely with interest rates in other countries, especially with the U.S. Federal rate. The Singapore Overnight Rate Average (SORA)[2] is a weighted average rate of all unsecured Singapore dollar overnight cash transactions brokered in Singapore between 9 am and 6:15 pm. In contrast with the Singapore Interbank Overnight Rate (SIBOR), which is denominated in

[2] The definition is from the Monetary Authority of Singapore (MAS). More information is available at http://www.sgs.gov.sg/sgs_data/daily_domestic_interbank_rates.html.

the U.S. dollar, the SORA is updated daily to help enhance the transparency of the Singapore dollar money market. We use the daily change of the SORA as an indicator to study the effect of the interest rate on the Singapore stock market,

$$\mathrm{SORA}(n) = 100 \times \frac{i_s(n) - i_s(n-1)}{i_s(n-1)} \tag{5.14}$$

where $i_s(n)$ is the daily value of the SORA.

2. *International Stock Market Indicator (ISMI):*
A report from the Monetary Authority of Singapore (MAS) supported a high level of the financial market integration and assessed the extent to which the openness of the Singapore financial market and capital flows attribute [94]. Koh and Wu [78] provided an analysis on the determinants of the Singapore stock exchange (SGX) during the 1997–1998 financial crisis, in which the biggest influence was from foreign stock exchanges. Major stock indices from the American, European and Asian markets have been investigated in the pre-crisis and the crisis periods, where the American market showed the closest relations with the Straits Times Industrials Index (STII) of Singapore, which was replaced by the STI from August 31, 1998. Suggested by these findings, we select the S&P 500 index from the American market, the HSI from the Asian market and the Financial Times and Stock Exchange (FTSE) 100 Index[3] from the European market as the potential factors to measure the influences of the international markets on the SGX. The internal residues of these three indices are obtained through the framework with the OE models given, respectively, by (5.3), (5.4) and

$$H_{\mathrm{FTSE}}(z) = \begin{bmatrix} \dfrac{4.971z^{-1} - 2.899z^{-2} + 1.319z^{-3} - 0.0397z^{-4}}{1 - 0.5939z^{-1} + 0.1773z^{-2}} \\ \dfrac{-2.098z^{-1} - 0.4305z^{-2} + 0.6319z^{-3} + 0.7053z^{-4}}{1 + 0.1122z^{-1} - 0.7651z^{-2}} \\ \dfrac{0.6217z^{-1} - 1.117z^{-2} - 0.9837z^{-3} + 0.7286z^{-4}}{1 - 0.3188z^{-1} - 0.1161z^{-2}} \end{bmatrix}^{\mathrm{T}}. \tag{5.15}$$

3. *Exchange Rate of the Singapore Dollar against the U.S. Dollar (SGD/USD):*
The exchange rate is investigated in almost all the literature concerning factors that affect the Singapore stock market. Since Singapore has a small, open and export-oriented economy, the exchange rate of the Singapore dollar is considered to have a significant impact on its economy and consequently on its stock market. Maysami and Koh [95] suggested that the Singapore stock market is sensitive to the interest rates and exchange rates on the macroeconomic side. The Singapore dollar against the currencies of certain developed countries were tested to be negatively related to the stock prices by Wu [130]. The results obtained by

[3] The STI is the headline index of the FTSE ST Index Series, which comprises a total of 77 indices.

Koh and Wu [78] and Yu [139] were a little different. They found that the foreign exchange market did not have predictive power over the STII before the 1997–1998 financial crisis. However, during the crisis period, it Granger caused the STII. The Singapore dollar is managed against a basket of the currencies of the major trading partners and competitors of Singapore. The composition of the basket is undisclosed and revised periodically by the MAS, but the U.S. dollar is commonly expected to share the largest proportion in the basket. We thus adopt the SGD/USD currency pair as the exchange rate indicator. Again, the change in the SGD/USD exchange rate is calculated in a similar way as that in (5.14).

4. *Oil Price (OP):*
   Since Singapore is a net oil importer, its economy is found to be vulnerable to oil price disturbances [71]. As to its stock market, the relationship is still controversial in the literature. Le and Change [83] concluded that it took time for the impact of oil price shocks to happen and work through the whole system in Singapore. However, considering that Singapore is in a similar position as Hong Kong, we choose to investigate the effect of the oil price indicator on the STI, as studied in the U.S. and the Hong Kong markets.
5. *Baltic Dry Index (BDI):*
   In recent years, Singapore has become one of the busiest ports in the world. About 7% of Singapore's GDP is contributed by its maritime industry. As the BDI was found to be an influential factor of the HSI, it is a natural selection for the STI as well. However, as we will find out shortly, the BDI does not seem to be having much impact on the STI. It will thus be discarded from the final set of the influential factors for the STI.

5.4.2 Causality Tests

As usual, data adjustment and data normalization are conducted for the input factors selected before one can carry out meaningful causality tests. The selection of the lag length and initial values are similar to those for other markets. For the internal residues of the S&P 500, the HSI and the FTSE, the lag length is chosen to be 4. For other influential factors, it is set to be 10. The resulting linear causality testing results are presented in Figures 5.19 – 5.25. We observe that only the HSI shows significant linear causal effect to the internal residue of the STI over the entire period of interest. The significant linear causality does not exist in the S&P 500 from January 24, 2006 to July 19, 2006. The same situation happens to the SORA from January 24, 2006 to January 23, 2008, to the OP from January 24, 2006 to November 29, 2007, and to the SGD/USD from January 24, 2006 to July 20, 2006. No significant linear causality exists in the FTSE and the BDI over the whole period. We then further carry out the nonlinear causality tests for the indicators during the periods when there is no linear causality observed. The results are given in Table 5.13. The nonlinear causality from the SGD/USD and FTSE to the internal residue of the STI are found to be significant, whereas the others are not.

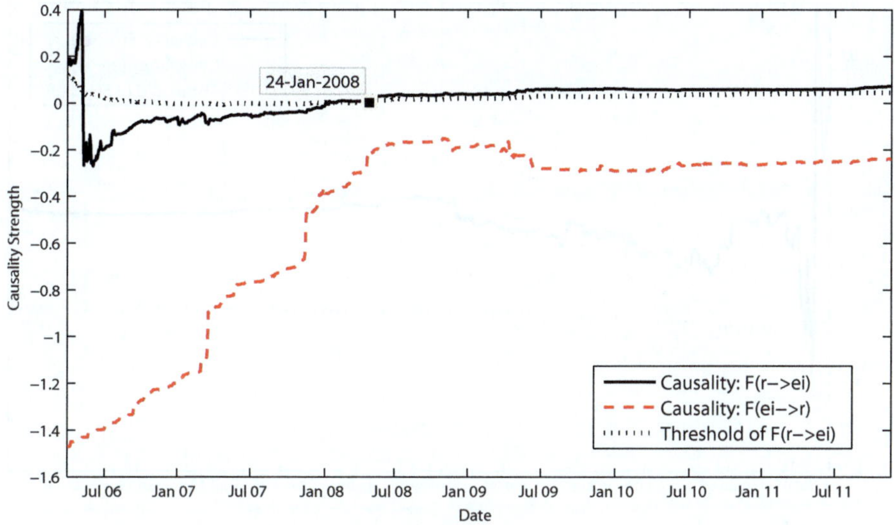

Fig. 5.19 Time-varying causality between the internal residue of the STI and the SORA

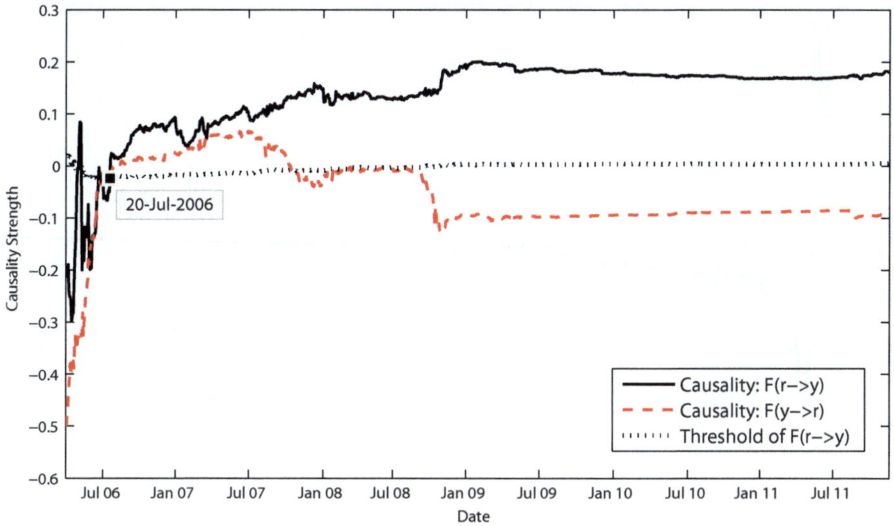

Fig. 5.20 Time-varying causality between the internal residues of the STI and S&P 500

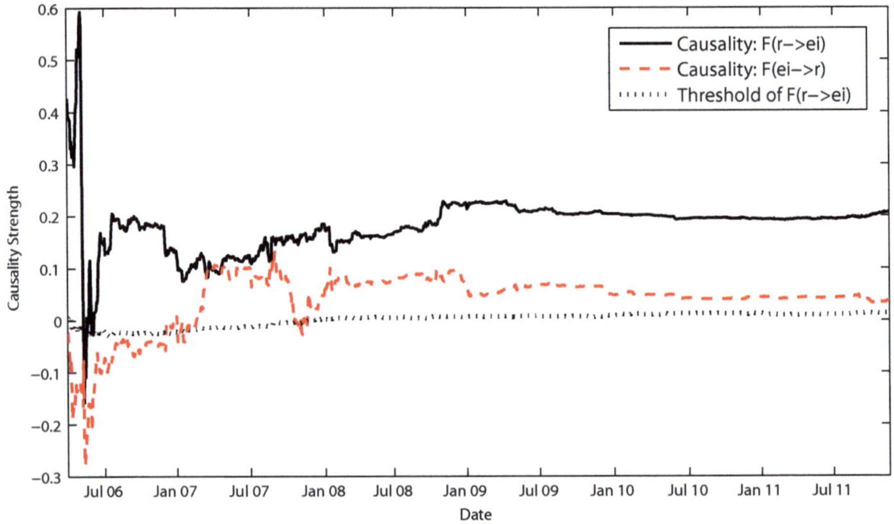

Fig. 5.21 Time-varying causality between the internal residues of the STI and HSI

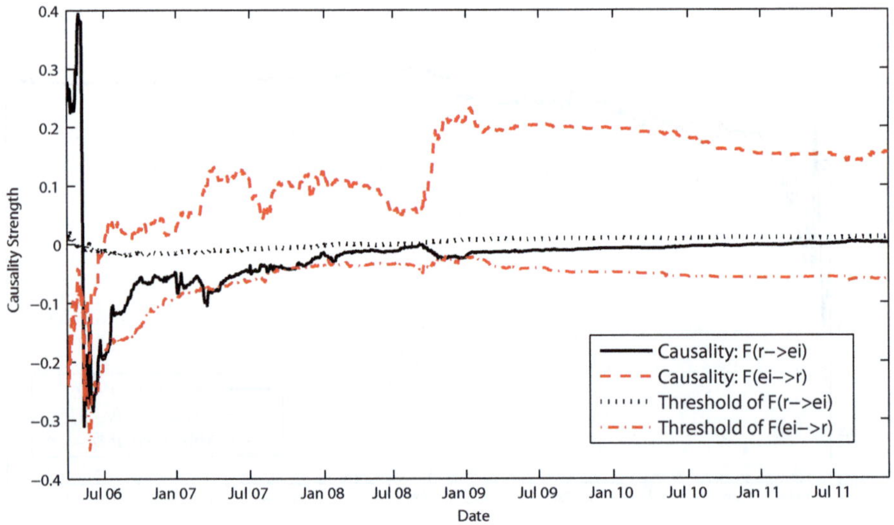

Fig. 5.22 Time-varying causality between the internal residues of the STI and FTSE

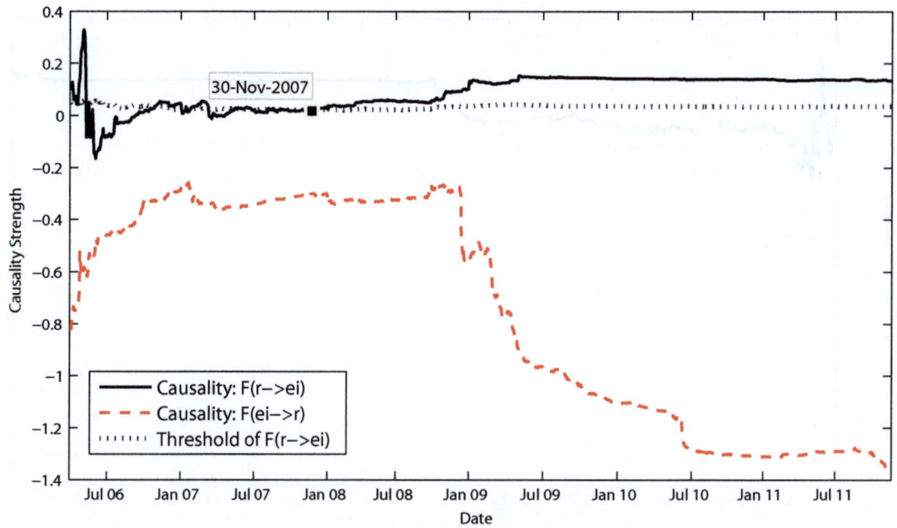

Fig. 5.23 Time-varying causality between the internal residue of the STI and the OP

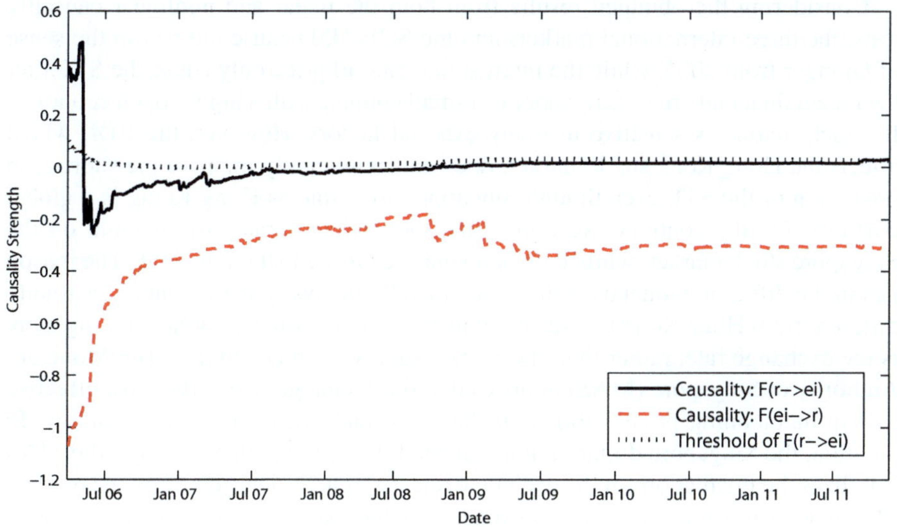

Fig. 5.24 Time-varying causality between the internal residue of the STI and the BDI

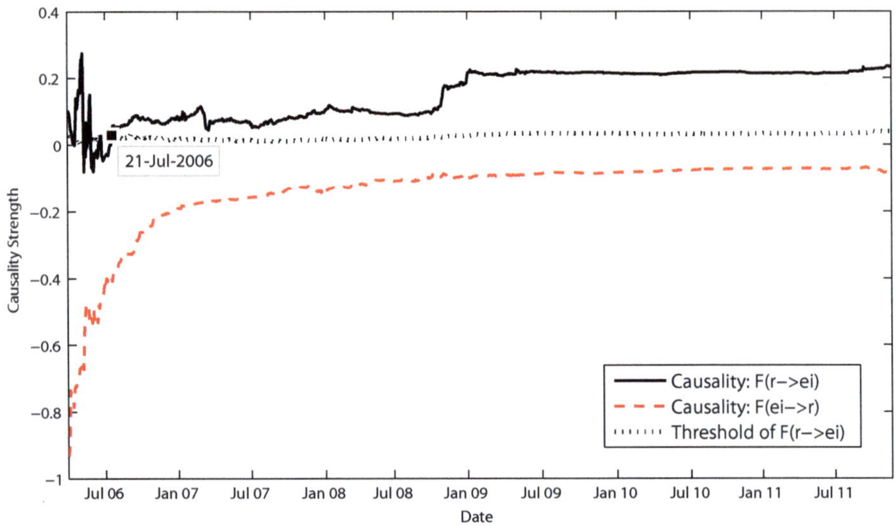

Fig. 5.25 Time-varying causality between the internal residue of the STI and the SGD/USD

Considering the obtained results from both the linear and nonlinear causality tests, the three international markets and the SGD/USD cause the STI in the sense of Granger from 2006, while the interest rate and oil price only cause the STI after the recent financial crisis. Singapore is a small country with a highly open economy. Its stock market is sensitive to many external factors. However, the BDI, which affects the Hong Kong stock market, does not significantly improve the short-term prediction of the STI, even though Singapore, the same as Hong Kong, is a global port city. On the contrary, we notice that the exchange rate is influential on the Singapore stock market, while it does not have causal effect on the HSI. The reason lies in the different monetary policies adopted by the two governments. It is totally different from Hong Kong that the main instrument of monetary policy in Singapore is the exchange rate, rather than the money supply or interest rates. The Monetary Authority of Singapore (MAS) believes that the exchange rate is the most effective tool in maintaining price stability in the small and open Singapore economy. In addition, the Singapore dollar is in a managed float regime that although the MAS stabilizes the fluctuations of the exchange rate, its trend is determined by the market. This unique monetary policy determines that the exchange rate affects the stock market of Singapore much more than the interest rate.

Since the causal effect from the S&P 500 began on July 20, 2006, we perform the multicollinearity test and one-step-ahead prediction starting from August 2006. According to the causality testing results, two subperiods are defined in the multicollinearity test, in which different influential factors are tested. It was from December 2007 and February 2008 when the oil price and interest rate, respectively, started to Granger cause the internal residue of the STI. Hence, the multicollinearity is calculated in two segments: (i) from August 2006 to January 2008 with the S&P

Table 5.13 Nonlinear Granger causality test results in the Singapore stock market

	S&P 500 does not cause STI internal residue (Jan. 24, 06–Jul. 19, 06)		FTSE does not cause STI internal residue		SORA does not cause STI internal residue (Jan. 24, 06–Jan. 23, 08)	
$Le = Lr$	CS	TVAL	CS	TVAL	CS	TVAL
1	−0.0024	−0.5595	0.0046	2.7846**	-2.88×10^{-4}	−1.8465
2	−0.0077	−0.8548	0.0080	3.1116**	-3.68×10^{-4}	−0.0068
3	−0.0223	−1.4698	0.0130	3.5227**	0.0031	0.7641
4	−0.0110	−0.5187	0.0195	3.9548**	0.0031	0.6816
5	−0.0080	−0.2504	0.0246	4.5372**	0.0023	0.5916
6	0.0094	0.1890	0.0320	5.1215**	0.0018	0.4597
7	−0.1289	−1.3679	0.0394	5.7215**	0.0028	0.7217
8	−0.1944	−1.0730	0.0441	6.0394**	0.0025	0.6204
9	−0.3750	−1.9365	0.0545	6.7867**	0.0035	0.7327
10	NA	NA	0.0594	6.4713**	0.0041	0.7583

	OP does not cause STI internal residue (Jan. 24, 06–Nov. 29, 07)		BDI does not cause STI internal residue		SGD/USD does not cause STI internal residue (Jan. 24, 06–Jul. 20, 06)	
$Le = Lr$	CS	TVAL	CS	TVAL	CS	TVAL
1	-9.25×10^{-5}	−0.0969	8.57×10^{-4}	0.8191	-3.63×10^{-4}	−0.9952
2	−0.0012	−0.5081	7.97×10^{-4}	0.4779	-6.84×10^{-5}	−0.0791
3	−0.0048	−1.0837	0.0034	1.0884	0.0038	1.3902*
4	−0.0035	−0.6186	0.0038	0.9531	0.0075	1.7756**
5	−0.0026	−0.3392	0.0038	0.8293	0.0105	1.8365**
6	−0.0040	−0.3632	0.0056	1.0166	0.0227	2.0447**
7	−0.0104	−0.6848	0.0104	1.5848*	0.0296	2.0901**
8	−0.0084	−0.4922	0.0130	1.6928**	0.0392	2.3535**
9	−0.0149	−0.7370	0.0111	1.2896*	0.0432	2.3731**
10	−0.0111	−0.3440	0.0101	1.0656	0.0444	2.4098**

* Significance at 10% level for a one-sided test.
** Significance at 5% level for a one-sided test.

500, the HSI, the FTSE and the SGD/USD; and (ii) from February 2008 to November 2011 with the S&P 500, the HSI, the FTSE, the SGD/USD, the SORA and the OP. Table 5.14 shows all the resulting tolerances, which are all close to 1, indicating little multicollinearity among the influential factors considered. As such, all these indicators are to be used as the input to the adaptation framework of the Singapore market, however, with the SORA and the OP being considered only in the period from February 2008 to November 2011.

Table 5.14 Multicollinearity test results in the Singapore stock market

	Tolerance	
	Aug. 2006–Jan. 2008	Feb. 2008–Nov. 2011
S&P 500	0.812	0.781
HSI	0.938	0.814
FTSE	0.885	0.861
SGD/USD	0.952	0.833
SORA	–	0.997
OP	–	0.811

5.4.3 Predicting Performance and Analysis

Similar to the Hong Kong market, five subperiods are defined from August 2006 to November 2011. The first subperiod (S1) is selected from July 2007 to January 2008, i.e., the first phase of the 2007 financial crisis. Its corresponding training data set is from August 2006 to June 2007, i.e., the pre-crisis period. When the global financial crisis started to spread from September 2008, the STI experienced the worst crash in its history. Although there was a fluctuation, the performance of the STI was relatively steady from February 2008 to August 2008. After September 2008, the stock market approached to a dangerous stage. Thus, we set the second subperiod (S2) from February 2008 to August 2008, a relatively steady stage during the crisis but with lots of bearish signs. From this subperiod, the interest rate and the oil price are considered as a part of the input to our framework. The most crazy period characterized by collapses and panic is selected as the third subperiod (S3), which is from September 2008 to February 2009. As illustrated in Figure 5.26, a rally of the STI started from March 2009. As it will be reported later in Chapter 6, there will be a market turning period in the middle of 2011, and the variance of the internal residue gets larger in 2011, the fourth subperiod (S4) is thus selected to be from March 2009 to the end of 2010. Finally, the last subperiod (S5) is chosen as the interval from January 2011 to November 2011. As shown in Figure 5.27, in every subperiod, there is at least one cluster of the internal residue with large variance.

The training data used for obtaining the corresponding adaptive filter for the five subperiods are illustrated in Table 5.15. The selection criteria of each training set are similar to those used in the Hong Kong market. The length of the training data is selected between half year to one year, depending on the available data of the influential factors. The initial values of the hyperparameters in the adaptive filter (the diagonal entries of Q_r) for different subperiods are also given in Table 5.15. Other initial conditions are the same as those in the other markets.

The comparison of predicting performances with different training data are given in Tables 5.16 – 5.18. More specifically, in Tables 5.16 and 5.17, we demonstrate that the predicting results can be enhanced by using more training data, while we show in Table 5.18 that the prediction is more accurate when new influential factors are added into the framework.

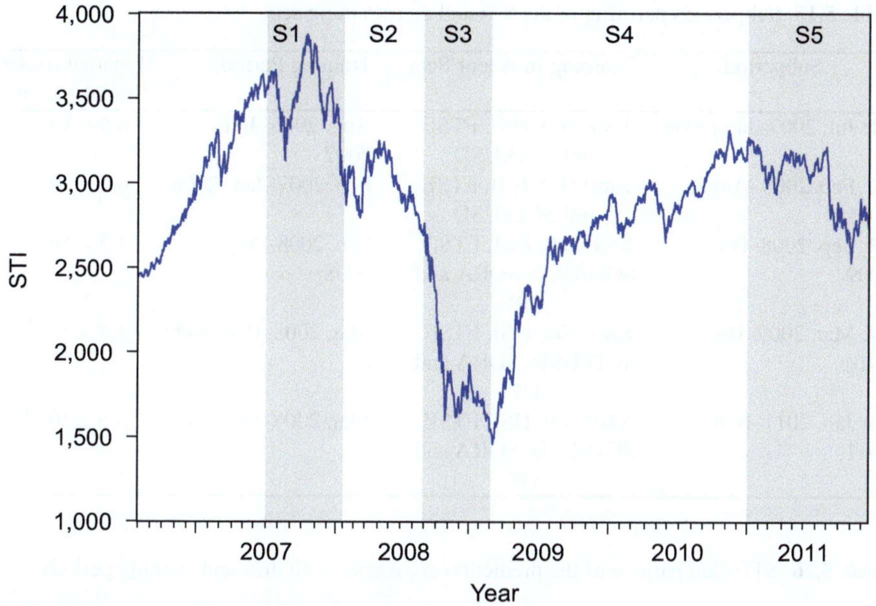

Fig. 5.26 Daily closing prices of the STI from August 2006 to November 2011

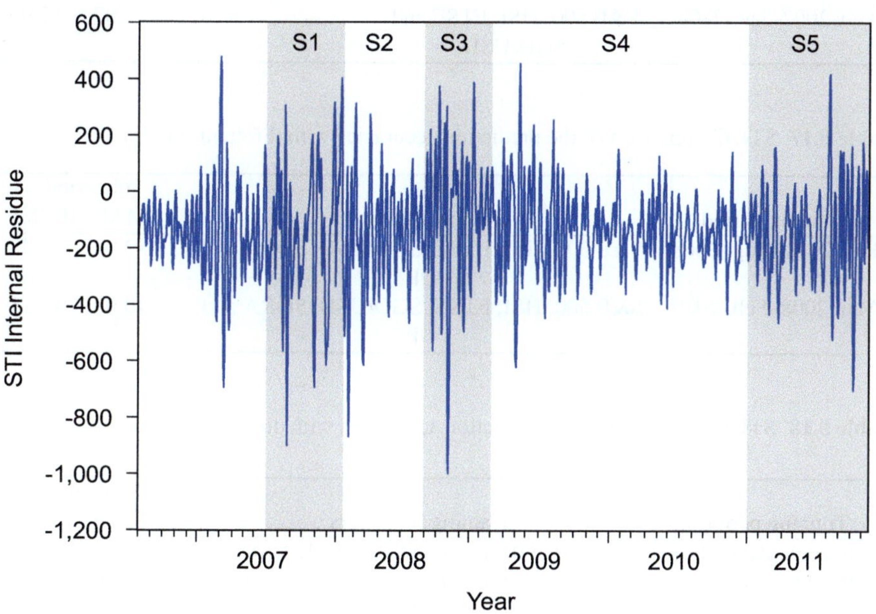

Fig. 5.27 Internal residue of the STI from August 2006 to November 2011

Table 5.15 Subperiods partition of the STI and their training sets

Subperiod	Training Indicator Set	Training Period	Hyperparameter Initial Values
S1: Jul. 2007–Jan. 2008	S&P 500, HSI, FTSE and SGD/USD	Aug. 2006–Jun. 2007	6.5×10^{-3}
S2: Feb. 2008–Aug. 2008	S&P 500, HSI, FTSE and SGD/USD	Feb. 2007–Jan. 2008	6.5×10^{-3}
S3: Sep. 2008–Feb. 2009	S&P 500, HSI, FTSE, SGD/USD, SORA and OP	Feb. 2008–Aug. 2008	3.5×10^{-4}
S4: Mar. 2009–Dec. 2010	S&P 500, HSI, FTSE, SGD/USD, SORA and OP	Mar. 2008–Feb. 2009	3.5×10^{-4}
S5: Jan. 2011–Nov. 2011	S&P 500, HSI, FTSE, SGD/USD, SORA and OP	Mar. 2008–Feb. 2009	3.5×10^{-4}

Table 5.16 STI: Comparison of the prediction accuracies with different training periods

		Subperiod S1		Subperiod S2	
Training period	Training indicators	MAE	RMSE	MAE	RMSE
Aug. 2006–Jun 2007	S&P 500, HSI, FTSE and SGD/USD	20.67	29.65	17.43	27.09
Feb. 2007–Jan. 2008	S&P 500, HSI, FTSE and SGD/USD	–	–	12.87	17.96

Table 5.17 STI: Comparison of the prediction accuracies with different training periods

		Subperiod S4	
Training period	Training indicators	MAE	RMSE
Feb. 2008–Aug. 2008	S&P 500, HSI, FTSE, SGD/USD, SORA and OP	12.95	17.48
Mar. 2008–Feb. 2009	S&P 500, HSI, FTSE, SGD/USD, SORA and OP	10.08	14.15

Table 5.18 STI: Comparison of the prediction accuracies with different training indicator set

		Subperiod S3	
Training period	Training indicators	MAE	RMSE
Feb. 2008–Aug. 2008	S&P 500, HSI, FTSE and SGD/USD	27.40	39.13
Feb. 2008–Aug. 2008	S&P 500, HSI, FTSE, SGD/USD, SORA and OP	12.50	15.78

Finally, we summarize the one-step-ahead predicting performances for the Singapore stock market in Table 5.19. As usual, for the ARMAX model, the lag length is set to be 4 for the AR and MA terms, and 10 for all the other exogenous inputs. For references, we note that the average daily changes of closing prices in these five subperiods are, respectively, 44.29, 31.62, 43.84, 23.35 and 27.14. It is clear that the proposed system adaptation framework has shown its superiority once again in predicting the market movement.

Table 5.19 Comparison of prediction accuracies between the ARMAX approach and the system adaptation framework for the STI

	Subperiod S1		Subperiod S2		Subperiod S3	
	MAE	RMSE	MAE	RMSE	MAE	RMSE
ARMAX	40.53	55.42	30.07	39.94	43.19	56.19
System adaptation framework	20.67	29.65	12.87	17.96	12.50	15.78
Improvement (%)	49.00	46.50	57.20	53.03	71.06	71.92

	Subperiod S4		Subperiod S5	
	MAE	RMSE	MAE	RMSE
ARMAX	21.99	29.78	24.29	31.94
System adaptation framework	10.08	14.15	9.79	13.46
Improvement (%)	54.16	52.48	59.70	57.86

Chapter 6
Market Turning Period Forecasting

6.1 Introduction

In this chapter, we adopt the system adaptation framework to forecast major market turning periods, which can be roughly defined as a transitional period that the market is turning either from a bullish phase to a bearish trend or vice versa. Many factors have been used for forecasting the major turnings in stock markets. One of the most widely used factors is the business cycle. It always gives a direct implication of major market turnings. The concept of the business cycle was introduced by Samuelson's multiplier accelerator model [111], referring to the periodic but irregular fluctuation in economic activities. It is considered as an indicator of market turnings since it is closely related to the periods of economic boom and recession. Using time-frequency analysis, Chen [30, 31] proved the existence of persistent chaotic cycles in the U.S. stock market. In [30, 31], the analysis of the relationship between frequency patterns and dynamic changes in the business cycle provided an explanation to the U.S. stock market crash in October 1987. Other important economic indicators frequently considered include (i) quantitative indicators, e.g., the price/earnings ratio, dividend yield ratio, inflation rate and other macroeconomic variables, and (ii) qualitative indicators, e.g., the leverage level of shadow banking system, general situations of mergers and acquisitions and central bank liquidity injection.

It is believed that in the long term, the stock market prices are influenced by macroeconomic factors. As such, the macroeconomic factors are widely used to judge whether or not the stock market is undergoing a correction or a major turning. Besides the interest rate, the most well-known macroeconomic factor, a substantial body of the literature considers the inflation rate to be another macroeconomic factor that seriously affects the market. For example, Niemira [101] studied the relationship between the inflation rate and stock market cycles, and suggested using it as an indicator to predict the turning points in the stock market. Golob and Bishop [54] found that the U.S. stock prices over thirty years from the mid-1970s closely followed its inflation rates. This relationship in different timescales was also evidenced by Kim and In [76] using wavelet analysis. When the structural breaks of the inter-

X. Zheng & B.M. Chen: *Stock Market Modeling and Forecasting*, LNCIS 442, pp. 91–122.
DOI: 10.1007/978-1-4471-5155-5_6

est rate and the inflation rate appear, they usually coincide with important economic events such as structural changes in the financial market and financial crisis. The reaction of the stock market is a sharp crash or a rally.

However, in the technical analysis approach, the market trend and possible turnings are detected merely based on stock prices alone without actually considering the above-mentioned indicators. In technical analysis, it is assumed that if a trend has been established, the future price movement is more likely to follow the same direction of the trend. By analyzing the patterns of the historical prices, the technical analysis is able to provide simple but powerful tools to identify the market trend such as the trend line, support and resistance, etc. Inspired by this idea that stock prices can reflect most of the useful information of the long-term market trend, we choose to forecast the major market turnings mainly using stock prices. Since we are going to focus on the long-term market trends, only the properties and signals related to the internal model of the proposed system adaptation framework are to be used in forecasting the market turning periods. More specifically, we present in Section 6.2 a technique for predicting the market turning period by transforming the time series of the internal residue into a frequency domain response using the fast Fourier transform (FFT), while in Section 6.3, we propose an alternative confirmation method based on the stability properties of the internal OE model. We will test these techniques on the DJIA, the SSE, the HSI and the STI.

6.2 A Frequency Domain Forecasting Technique

When a long-term market trend alters its direction, there will be some essential changes in the market internal dynamics. In our framework, such internal dynamics are characterized by the internal model I, while the difference between the output of the internal model and the actual stock price is defined as the internal residue. Thus, to reveal the essential market properties, we proceed to analyze the internal residue in the frequency domain using the well-known fast Fourier transform (FFT). We should note that the FFT is a very common technique, which is heavily utilized in analyzing and solving engineering problems. It is capable of providing some unique frequency domain properties of the signal or system, which cannot be captured in the time domain. The analysis of financial time series in the frequency domain is a relatively underexplored area in the literature.

Interestingly, we have found that frequency contents of the internal residue vary a lot from time to time. Some significant components in the power spectrum of the internal residue can be obviously observed (see, e.g., Figure 6.1). Tests find that the sudden change in the power spectrum always signifies a major turning in the stock market. To facilitate the analysis, the time span is chosen as multiples of months, but in practice, it would be updated every day. More specifically, we perform the FFT on the sampled data of the internal residue until a unique frequency pattern with significant peaks appears and disappears, and thus identify a possible market turning period. To find out the next turning period, we use the same procedure but with a new starting sample point. In order to avoid the drastic fluctuation during the transition of two trends, we select the new starting point at about four to five months after the previously identified turning period. This time interval is also used to confirm whether or not the identified period is indeed corresponding to a major market turning. If it is proven to be incorrect, we continue the forecasting procedure of the earlier part. If the new trend is confirmed, we adjust the new starting sample point accordingly. Even though such a procedure is purely empirical, it has been proven to work quite well in the real stock market.

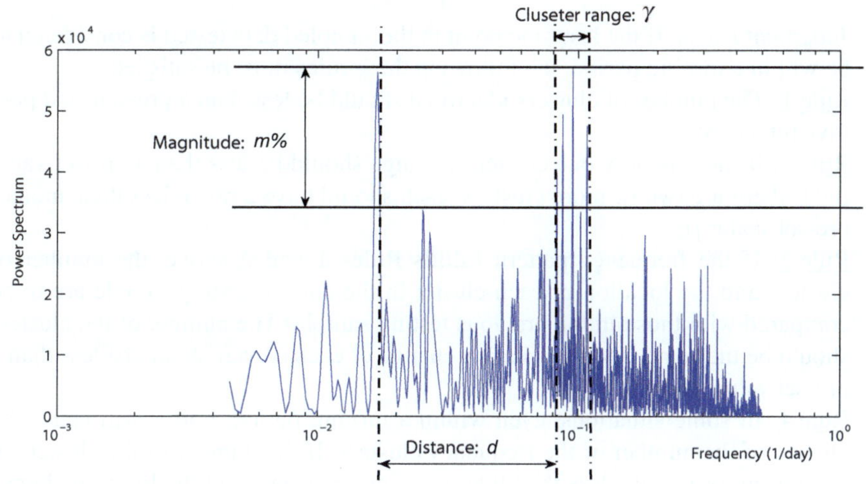

Fig. 6.1 Elements in frequency domain identification

Shown in Figure 6.1 is a typical frequency pattern of the internal residue generated by the system adaptation framework for the DJIA. In order to realize the turning period forecasting technique using computer software, we develop a set of empirical rules to recognize frequency patterns for the turning periods. Our rules concern the magnitude, the distance between frequency components and the range and location of a cluster in the frequency components, which are, respectively, illustrated in Figure 6.1. For the power spectrum of the internal residue, we first extract its necessary patterns and features and then make a judgment on weather or not there is a characteristic change in its frequency components (i.e., there is a market turning period occured).

The searching procedure for the market turning periods goes as follows:

1. *Feature Extraction:*
   <u>Step 1</u>: Given a pre-set positive scalar m, if the magnitude of the frequency response is greater than $m\%$ of the maximum peak, we retain it for further processing. Otherwise, it is to be removed. This step repeats for all the frequency points considered.

Step 2: The distance between two adjacent frequency components, d, is defined as their frequency differences in a logarithmic scale. If $d > \iota$, where ι is a pre-set positive scalar, we categorize these two frequency components into different clusters. Otherwise, they are considered to be within the same cluster.
Step 3: The range of a cluster, γ, is defined as the logarithmic distance between the first and the last frequency components in the cluster. The location of a cluster is defined as the frequency of the component, which has the largest magnitude in the cluster.

2. *Judgment Rules:* If the last time point in the sampled data tested is considered to be within a turning period, the following three rules must be satisfied:
   Rule 1: The number of clusters identified should be less than a pre-selected positive integer κ.
   Rule 2: If there is only one cluster, its range should be less than a pre-set value ρ_1. If there are two or more clusters, each should have a range less than another pre-set scalar ρ_2.
   Rule 3: If the frequency pattern fulfills Rules 1 and 2 above, the number of clusters and the location of each cluster in the current testing sample are to be compared with those in the previous testing sample. The number of the clusters should be the same, and the location change of each cluster should be less than a pre-set value ς.
   Rule 4: In some situations, even within a turning period, noises will cause the change of the number of the frequency clusters. If the number of the clusters in the current test is one less than that in the previous test and the location change of each cluster is still within ς, it is considered to be within the turning period.
   Rule 5: If an additional cluster appears, we will use the above rules to judge whether this new cluster has dominant components. Specifically, we will remove the original clusters of dominant components, and then use Rules 1 and 2 to re-judge the remaining components. If Rules 1 and 2 are fulfilled, the situation is considered to be changed as there are new dominant frequency components appeared. Otherwise, this new cluster is to be ignored. In this rejudgment, κ is set as 1 and the threshold values of m and ρ_1 are reset to another values, $\tilde{m}$ and $\tilde{\rho}_1$, respectively.

For the markets of interest, i.e., the DJIA, the SSE, the HSI and the STI, we find that the following threshold values work pretty well:

$$\iota = 0.2, \quad \kappa = 3, \quad \rho_1 = 0.23, \quad \rho_2 = 0.13, \quad \varsigma = 0.001, \quad \tilde{\rho}_1 = 0.15, \quad \tilde{m} = 75\% \tag{6.1}$$

and m is set to 60% for the DJIA, the SSE, the STI and 65% for the HSI.

We should note that the above procedure is just a preliminary result that we have used in computer software implementation. It is neither rigorous nor perfect. A possible solution could be to employ appropriate filters as those used in engineering systems to filter out unwanted components in the frequency response of the internal residue so that a rigorous procedure can be developed to identify the market turning periods. This could be another interesting topic for future research.

6.2.1 Dow Jones Industrial Average

In what follows, we proceed to forecast the market turning periods for the DJIA. It was reported in Chen [30, 31] that there are persistent chaotic cycles in the U.S. stock market, which were around three to five years. Therefore, it is sufficient to use a five-year data set for estimating the internal model. In order to capture new market trends, the internal model will be re-estimated by using a new set of market data once there are two turning periods (a pair of peak and bottom) found.

The daily closing prices of the DJIA from 1990 to 1994 are used to estimate the initial internal model, which gives

$$H_{\mathrm{oe,DJIA},1}(z) = \begin{bmatrix} \dfrac{-2.566z^{-1}+2.120z^{-2}-0.061z^{-3}-0.371z^{-4}}{1-0.197z^{-1}-0.110z^{-2}} \\ \dfrac{1.866z^{-1}-1.581z^{-2}-0.058z^{-3}+0.537^{-4}}{1-0.563z^{-1}+0.085z^{-2}} \\ \dfrac{1.707z^{-1}-2.033z^{-2}+2.263z^{-3}-1.338z^{-4}}{1-0.231z^{-1}-0.023z^{-2}} \end{bmatrix}^{\mathrm{T}} . \tag{6.2}$$

Shown in Figure 6.2 are the frequency responses of the resulting internal residue of the DJIA with progressive sample data. All the nine plots have the same starting point, i.e., January 3, 1995. The time interval for the first plot is 24 months, and it increases gradually for the subsequent plots to the last one, which has a sample size of 48 months in total. We can observe that there is one significant peak showing up in the fourth plot, in which its sample data terminate at the end of July 1999. It is considered to be the beginning of a turning period by our proposed rules. After this, although the subsequent sample size keeps increasing, the frequency response of the internal residue remains relatively similar. The pattern, however, disappears in the last plot, indicating the end of this turning period. By the rules, we consider that the DJIA has a turning period from August 2 to November 30 in 1999.

The next forecasting process begins at the sample point of May 1, 2000, five months after the previous turning period. The frequency responses of the internal residue are respectively shown in Figure 6.3, in which the interval from August 1 to December 31, 2002 is identified as another turning period. By the proposed rules, these two turning periods are verified and confirmed. In addition, during this period, two more turning periods, (i) from February 1 to March 30 in 2001, and (ii) from September 4 to December 5 in 2001, are detected by using our forecasting technique, which are marked in the red shaded areas in Figure 6.4. We note that even though they do not correspond to major market turnings, a sharp decline followed by a quick recovery can be observed in both incidents.

After the second turning has been detected, the internal OE model is to be re-estimated. We use the closing prices of the DJIA from January 3, 1995 to January 31, 2003 as training data and obtain the following internal model:

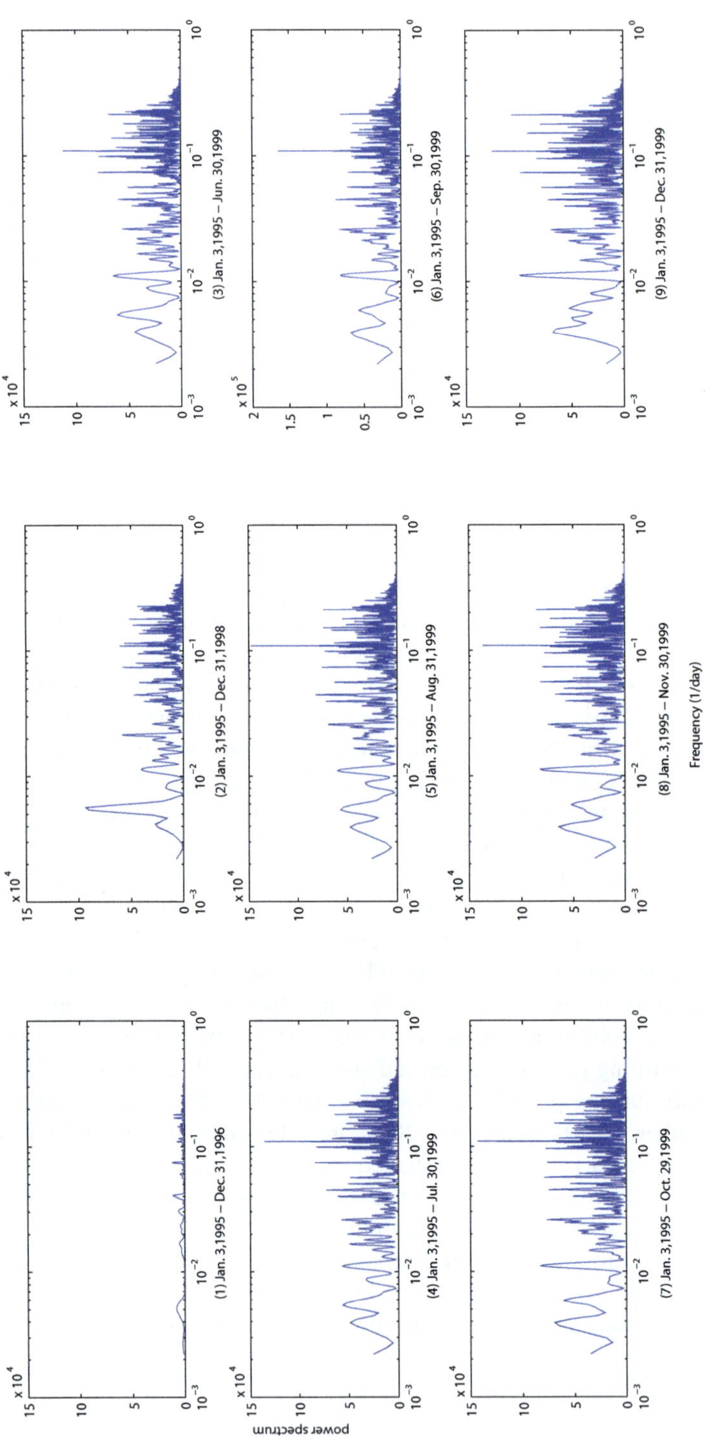

Fig. 6.2 Frequency responses of internal residue of the DJIA from January 1995 to December 1999

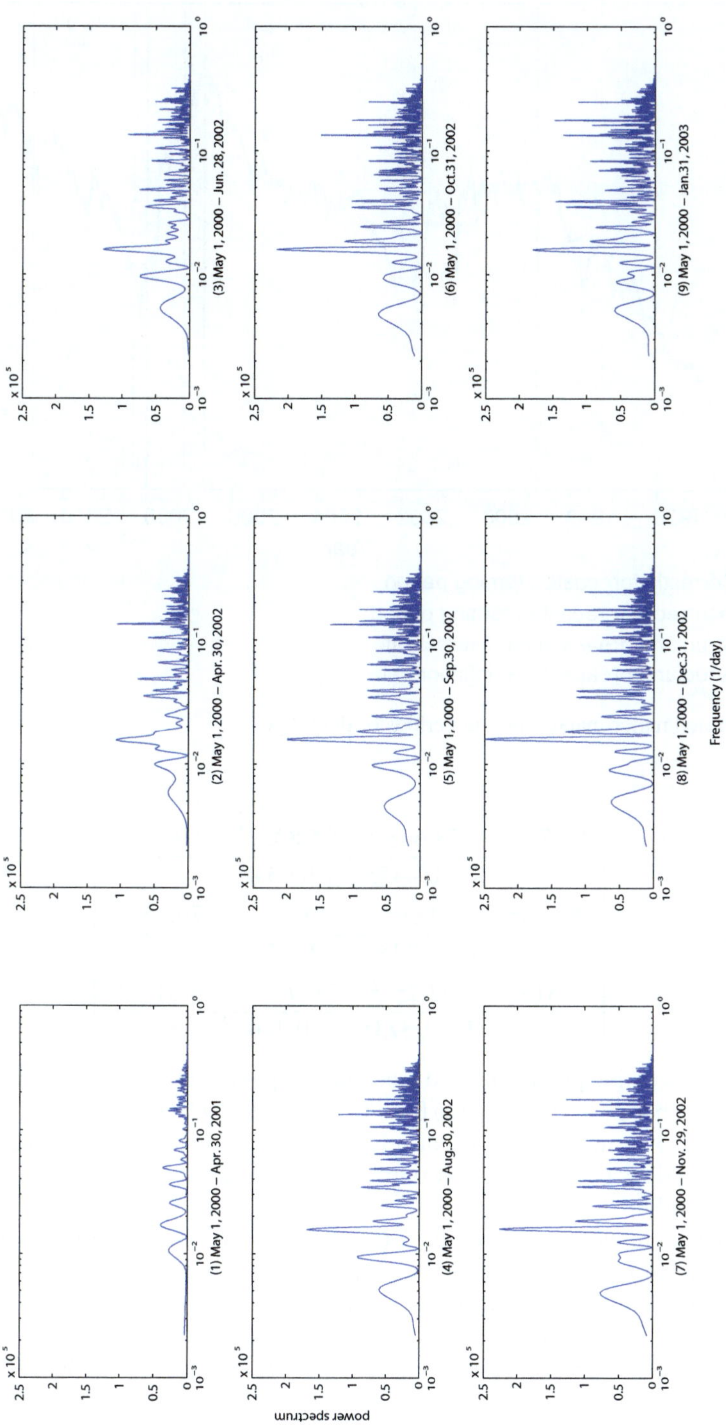

Fig. 6.3 Frequency responses of internal residue of the DJIA from May 2000 to January 2003

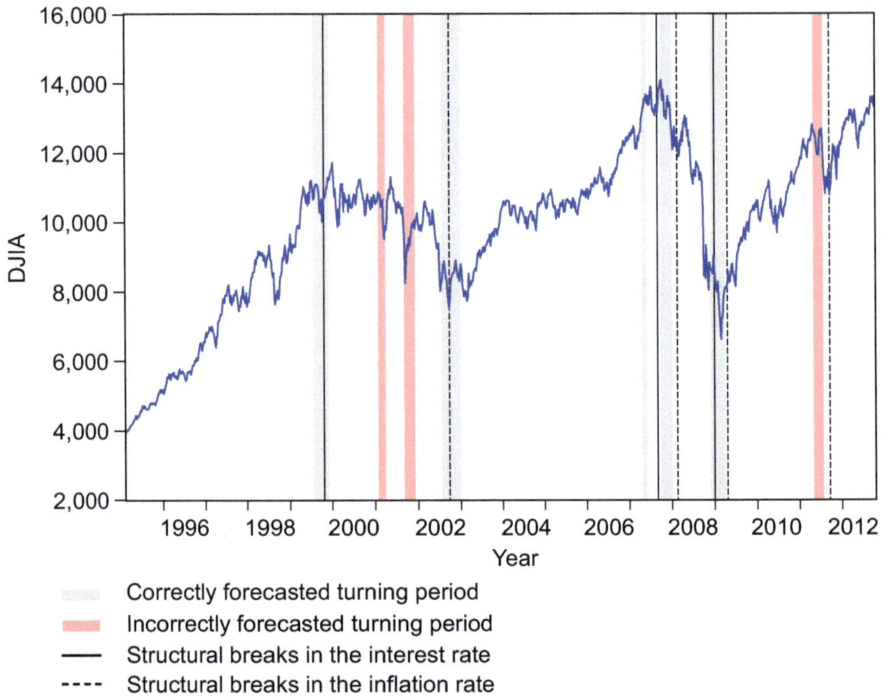

Fig. 6.4 Forecasted major market turning periods of the DJIA

$$H_{\text{oe,DJIA},2}(z) = \begin{bmatrix} \dfrac{66.59z^{-1} + 58.66z^{-2} - 25.88z^{-3} - 0.942z^{-4}}{1 - 0.445z^{-1} + 0.034z^{-2}} \\ \dfrac{52.18z^{-1} - 24.66z^{-2} - 151.1z^{-3} - 76.65^{-4}}{1 + 0.825z^{-1} - 0.120z^{-2}} \\ \dfrac{2.515z^{-1} - 57.2z^{-2} + 76.16z^{-3} - 25.01z^{-4}}{1 - 1.373z^{-1} + 0.446^{-2}} \end{bmatrix}^{\mathrm{T}}. \tag{6.3}$$

Following our forecasting procedure, two more turning periods are detected: (i) from April 25 to December 21 in 2007 and (ii) from December 1, 2008 to March 31, 2009. The results are shown in Figure 6.4. For references, the corresponding frequency responses used in detecting the above two turning periods are shown in Figures 6.5 and 6.6, respectively.

By contrast with movements of the DJIA (see Figure 6.4), it is found that most of the turning periods we have detected are correct. They are located in the transition periods between the bull and bear markets, i.e., the market reversals. The DJIA first closed above 11,000 on May 3, 1999 and reached a record high on January 14, 2000, which is close to the ending point of the first turning period detected. After this, the DJIA began to decline. Although there was a rally in the second half of

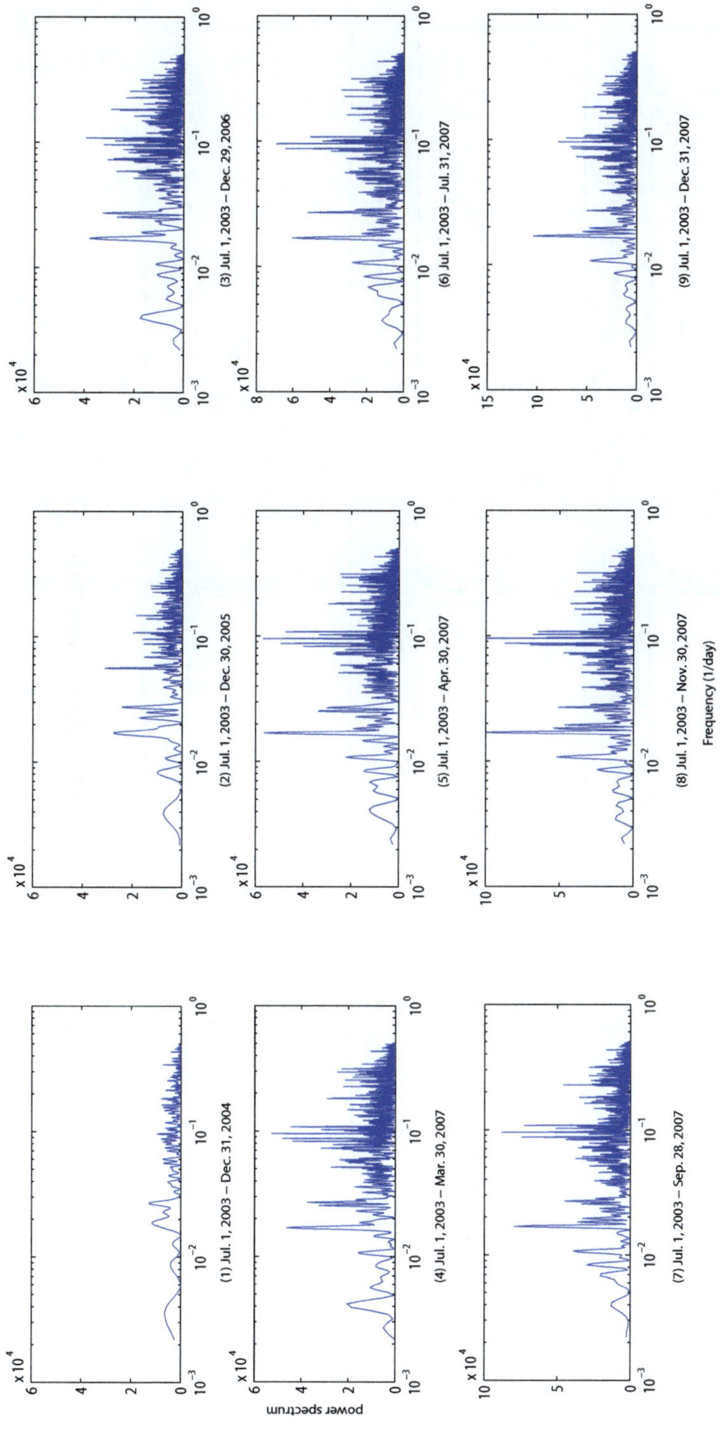

Fig. 6.5 Frequency responses of internal residue of the DJIA from July 2003 to December 2007

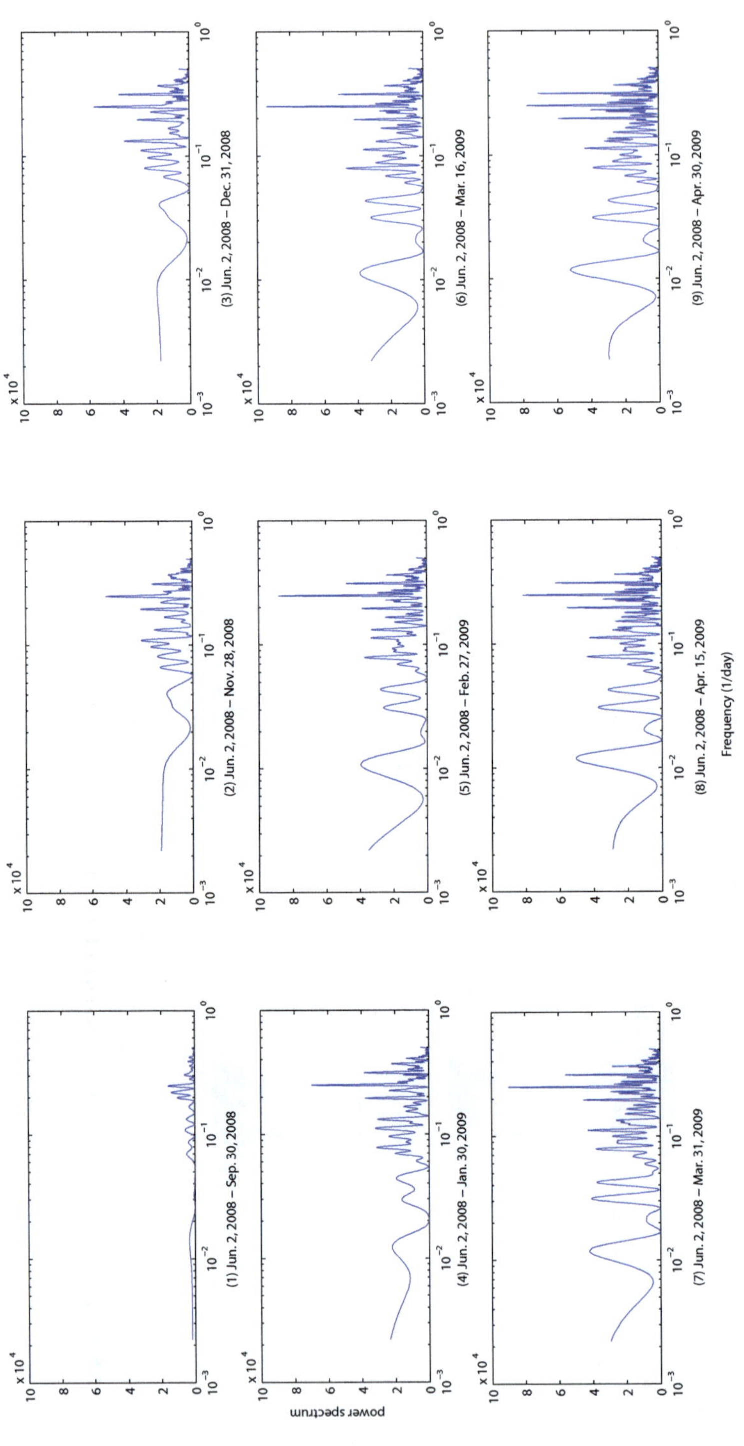

Fig. 6.6 Frequency responses of internal residue of the DJIA from June 2008 to April 2009

2000, it bottomed out at its lowest since October 1997 on October 9, 2002, which is considered as the ending of the three year bear market. From March 2003, the DJIA turned to recover slightly, and then bulls got overpowered again, which has also been detected by our method. After reaching the record high of 14,164.53 on October 9, 2007, it started to crash. That was when the 2007 subprime crisis, the latest and the most serious global financial crisis since the Great Depression, started. This corresponds to the third turning period. In the wake of this global financial crisis and ensuing stock market collapse in 2008, a series of financial support measures were launched to stimulate the economy. The fourth turning period we have forecasted matches the recovery of the DJIA from the beginning of 2009. After this turning period, the following internal model is obtained using a new set of training data:

$$H_{\text{oe,DJIA},3}(z) = \begin{bmatrix} \dfrac{0.0879z^{-1}+0.0392z^{-2}+0.0993z^{-3}-0.1389z^{-4}}{1-1.567z^{-1}+0.7318z^{-2}} \\ \dfrac{-0.0572z^{-1}-0.2309z^{-2}+0.134z^{-3}+0.0207^{-4}}{1-1.574z^{-1}+0.7255z^{-2}} \\ \dfrac{2.317z^{-1}-0.3122z^{-2}-0.2701z^{-3}+0.0111z^{-4}}{1+0.3505z^{-1}-0.0554^{-2}} \end{bmatrix}^{\mathrm{T}}. \tag{6.4}$$

With this OE model, a new turning period from May 2 to July 21 in 2011 can be detected. The DJIA experienced about 15% decline after this period. However, this turning period is marked in red as it is clearly not a major turning period. Up to October 2012, there is no other turning period found using our method.

As mentioned earlier, it is believed that the long-term trend of the stock market prices is influenced by the macroeconomic factors. In what follows, we investigate the linkage between the macroeconomic structural stability and the market turning periods that we have detected. We choose the interest rate (the 1-month time deposit interest rate) and the inflation rate as the macroeconomic indicators.

Most of the existing methods for detecting structural breaks are based on statistical tests. The classic approaches are the Chow test for a single known break [36], Andrew-Ploberger test for a single unknown break [4], and Bai-Perron test for multiple unknown breaks [8, 9]. We employ the well-developed Bai-Perron test to detect multiple structural breaks. In our studies, the AR model is adopted as the basis for the linear model regression. The optimal lag length is set based on the Akaike Information Criterion (AIC) with maximum ten lags and the results of the Durbin-Watson statistic and the Breusch-Godfrey serial correlation Lagrange multiplier test (B-G LM Test). The recommenced two-step testing strategy is adopted. The first step is to use a dual maximum test. It includes equal and different weighted versions, statistics of which are denoted by UDmax and WDmax, respectively. If UDmax and WDmax are significantly important, it indicates that there is at least one structural breaks in the series. A sequential test is then carried out to determine the number of the actual breaks. Otherwise, structural stability is considered. There is a trimming factor λ in the structural break test, defining a proportion of observations that is trimmed at each end. It is selected according to the length of the series in order to make sure that the observations before or after the break point are sufficient to estimate the regression relationship. Thus, the breaks found in the tests are always located in the middle of the segment.

We partition the whole time interval of interest into two segments: (i) from January 1995 to December 2010 and (ii) from January 2008 to October 2012. The test results for the DJIA are presented in Table 6.1. We note that the Durbin-Watson statistics are all close to 2 and the p values of the LM test are all larger than 0.05, indicating that there is no autocorrelation in the residuals with their optimal lags. The results of the dual maximum tests (UDmax and WDmax) are all significant at least at 5% level, revealing their structural instability. Therefore, the sequential test with upper bound at most 5 is conducted, i.e., we consider at most 5 structural breaks in each series. It provides the number and the locations of the breaks. For easy reference, we depict the locations of these structural breaks together with the forecasted turning periods in Figure 6.4. It is obvious that there are always structural breaks during the detected turning periods. However, we should also note that as the macroeconomic data are naturally delayed, it is not feasible to utilize the structural breaks to forecast market turning periods. Nevertheless, interested readers are referred to [141] for detailed explanation on the relationship among the structural breaks, the market turning periods and the U.S. economy.

6.2.2 China, Hong Kong and Singapore Markets

Next, we proceed to forecast the turning periods for the stock markets in China, Hong Kong and Singapore, i.e., the SSE, the HSI and the STI. The testing periods for the SSE and the HSI are both from January 2005 to October 2012, and it is from January 2003 to October 2012 for the STI. The structural stability tests are also carried out for the series of the interest rate and inflation rate of these economies. As in the U.S. market, the structural stability tests are also conducted in two segments for each of the above-mentioned three markets. The deposit interest rates in these three markets are all fixed for few months or even few years. Thus, we place more weight on the money market interest rates. In Hong Kong and Singapore, the monthly money market interest rates kept changing until early 2009, when it began to stay fixed for a relatively long period. As such, considering necessary observations for the trimming effect, the first testing segment is defined from the beginning of the testing period to the end of 2009, and the second segment is from the beginning of 2008 to August 2012 (the latest available data). For the Hong Kong and Singapore markets, the money market interest rates are tested in the first segment whereas the long-term bond yields (selected as the interest rate indicator) are tested in the second segment. For the China market, the money market interest rate is tested throughout these two segments as there is no long-term bond yield available.

Since all the forecasting technique and testing strategies and criteria are the same as those adopted in the U.S. market. We will omit most of the detailed procedures in the following subsections.

Table 6.1 Bai-Perron test results of structural breaks in the U.S. macroeconomy

Macroeconomic Series	Segment	Optimal Lag	Durbin-Watson stat.	B-G LM Test (p value)	UDmax	WDmax	$SupLR_T(l+1\|l)$		Break points
							$SupLR_T(2\|1)$	$SupLR_T(3\|2)$	
U.S. interest rate	1995 – 2010	4	2.02	0.17	26.17***	36.27***	21.94**	10.70	Oct., 1999 Aug., 2007
	2008 – 2012	3	2.01	0.43	23.21***	43.54***	12.37	–	Dec., 2008
U.S. inflation rate	1995 – 2010	3	1.99	0.37	16.63**	27.96***	17.38**	6.73	Sep., 2002 Feb., 2008
	2008 – 2012	4	2.00	0.11	15.89*	17.39*	18.24**	2.35	Apr., 2009 Sep.,2011

Note: * 10% level ** 5% level *** 1% level

6.2.2.1 Shanghai Stock Exchange Composite Index

The initial training period for the SSE is from January 2000 to December 2004, based on which the following internal OE model associated with the system adaptation framework is obtained:

$$H_{\mathrm{oe,SSE},1}(z)=\begin{bmatrix}\dfrac{2.242z^{-1}-1.774z^{-2}+0.734z^{-3}+0.130z^{-4}}{1-0.843z^{-1}+0.365z^{-2}}\\[2ex]\dfrac{-0.132z^{-1}-0.450z^{-2}+0.463z^{-3}+0.245^{z}-4}{1-0.209z^{-1}-0.090z^{-2}}\\[2ex]\dfrac{2.710z^{-1}-3.672z^{-2}-0.035z^{-3}+0.341z^{-4}}{1-0.739z^{-1}+0.119z^{-2}}\end{bmatrix}^{\mathrm{T}}. \tag{6.5}$$

Using the frequency domain technique, we obtain two turning periods, which are respectively (i) from August 15, 2007 to October 8, 2007 and (ii) from November 10, 2008 to May 13, 2009. The results are highlighted by the shaded areas in Figure 6.7. Both turning periods give an accurate and timely warning to a market reversal, but the second one lasts a little longer. After these two turning periods, we re-estimate the OE model using a new set of training data from January 2005 to May 2009. The resulting OE model is given as

$$H_{\mathrm{oe,SSE},2}(z)=\begin{bmatrix}\dfrac{0.3122z^{-1}-0.4729z^{-2}+0.5564z^{-3}-0.269z^{-4}}{1-1.73z^{-1}+0.9561z^{-2}}\\[2ex]\dfrac{1.704z^{-1}-2.709z^{-2}+0.8574z^{-3}+0.4762^{z}-4}{1-1.731z^{-1}+0.9552z^{-2}}\\[2ex]\dfrac{2.784z^{-1}-5.636z^{-2}+2.161z^{-3}+0.4871z^{-4}}{1-0.6872z^{-1}-0.1143z^{-2}}\end{bmatrix}^{\mathrm{T}}. \tag{6.6}$$

With the new OE model, a turning period from December 28, 2009 to February 9, 2010 is detected, indicating a decline in the market. Right after this turning period, a drastic slump of about 25% in the SSE occured from April 2010 to June 2010. However, the market had a mild rally after that. There is another turning period from June 29, 2010 to November 11, 2010, which we have found in the SSE. Comparing the locations and price levels of these two turning periods, it is rational to declare the first one (from December 28, 2009 to February 9, 2010) as an incorrect forecasting. Thus, it is marked in red in Figure 6.7. The second turning period (from June 29, 2010 to November 11, 2010) is then treated as a correct indication of a market downturn, which is proven to be correct as the trend of the SSE has indeed been going downwards since then.

The structural stability tests of the interest rate and inflation rate of China are given in Table 6.2 and the locations of the structural breaks are shown in Figure 6.7. As shown in Figure 6.7, the monthly money market interest rate is more indicative than the inflation rate for the China stock market. We note that for every turning period, there is a structural break in the series of the interest rate. For the four break points related to the inflation rate, only the one in December 2008 coincides

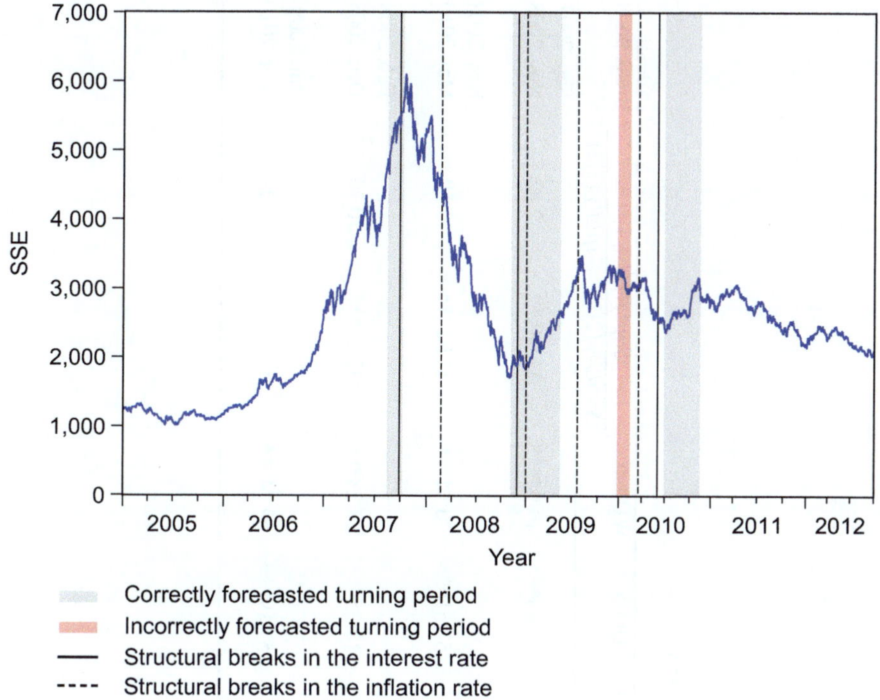

Fig. 6.7 Forecasted major market turning periods of the SSE

correctly with the turning period. Furthermore, the third break point of the inflation rate might be a reason for the second turning period to last longer than usual.

6.2.2.2 Hong Kong Hang Seng Index

The initial OE model for the system adaptation framework of the Hong Kong market obtained using the training data set from January 2003 to October 2012 is given as

$$H_{\text{oe,HSI},1}(z) = \begin{bmatrix} \dfrac{3.147z^{-1} - 1.398z^{-2} + 1.672z^{-3} - 0.1733z^{-4}}{1 - 0.4154z^{-1} + 0.4002z^{-2}} \\ \dfrac{-0.7784z^{-1} - 2.847z^{-2} + 0.1516z^{-3} + 0.7184^{z}-4}{1 + 0.058z^{-1} - 0.3141z^{-2}} \\ \dfrac{2.618z^{-1} - 3.868z^{-2} + 2.16z^{-3} - 0.3464z^{-4}}{1 - 1.145z^{-1} + 0.5459z^{-2}} \end{bmatrix}^{\mathrm{T}}. \tag{6.7}$$

As shown in Figure 6.8, a turning period from July 23, 2007 to October 10, 2007 is detected. Starting from October 2007, the HSI began its sharp decline and plunged more than 60% from its all-time peak. The next turning period, which consists of

Table 6.2 Bai-Perron test results of structural breaks in the China macroeconomy

Macroeconomic Series	Segment	Optimal Lag	Durbin-Watson stat.	B-G LM Test (p value)	UDmax	WDmax	$SupLR_T(l+1\|l)$		Break points
							$SupLR_T(2\|1)$	$SupLR_T(3\|2)$	
China interest rate	2005 – 2009	3	1.99	0.23	39.66**	49.94***	32.98***	2.82	Sep. 2007 Nov. 2008
	2008 – 2012	4	1.99	0.36	57.61***	70.24***	41.92***	11.60	Nov. 2008 May 2010
China inflation rate	2005 – 2009	4	1.98	0.69	17.35*	27.28***	32.45***	10.10	Feb. 2008 Dec. 2008
	2008 – 2012	4	2.01	0.89	57.19***	69.74***	57.12***	6.29	Jul. 2009 Mar. 2010

Note: * 10% level ** 5% level *** 1% level

two periods neck to neck from September 29, 2008 to December 23, 2008, successfully forecasts the following uptrend of the market in 2009.

As in the previous markets, after two turning periods are identified, a new OE model is to be re-estimated using an updated set of training data. We obtain the following OE model for the next phase:

$$H_{\text{oe,HSI},2}(z) = \begin{bmatrix} \dfrac{0.8719z^{-1} - 0.484z^{-2} - 0.583z^{-3} + 0.2807z^{-4}}{1 - 1.262z^{-1} + 0.33z^{-2}} \\ \dfrac{-2.283z^{-1} - 0.1264z^{-2} + 1.012z^{-3} + 0.8523^{z}-4}{1 - 0.6504z^{-1} - 0.221z^{-2}} \\ \dfrac{2.352z^{-1} - 0.9294z^{-2} - 0.3512z^{-3} + 0.0735z^{-4}}{1 - 0.9079z^{-1} + 0.1951z^{-2}} \end{bmatrix}^{\mathrm{T}}. \tag{6.8}$$

The resulting turning periods detected are shown in Figure 6.8.

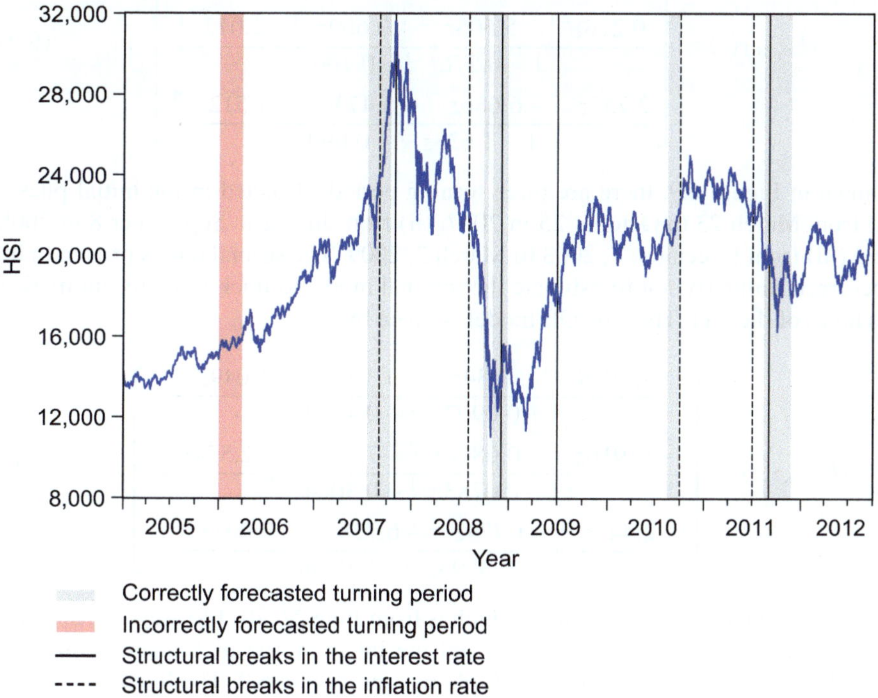

Fig. 6.8 Forecasted major market turning periods of the HSI

Testing results of the structural breaks in the Hong Kong macroeconomy are given in Table 6.3 as well as in Figure 6.8. Both the interest rate and the inflation rate show their structural breaks around the turning periods, giving another evidence of the connection between the internal model dynamics and the macroeconomic structural stability. However, we should note that there is a false turning period spotted in early 2006. We detect a turning period from August 16, 2010 to September 22, 2010, in which there is also a structural break in the inflation rate. After fluctuating for almost one year, the HSI tumbled about 25% from November 2010. Another turning period is found from August 15, 2011 to November 21, 2011, signifying a reversal from the beginning of 2012. Both the interest rate and the inflation rate present a structural break around this period.

6.2.2.3 Singapore Straits Times Index

For the STI, by using the daily close prices from January 1998 to December 2002, we obtain the following internal OE model for its system adaptation framework for forecasting the first two turning periods after 2003:

$$H_{\mathrm{oe,STI},1}(z) = \begin{bmatrix} \dfrac{5.701z^{-1}+3.797z^{-2}+1.257z^{-3}+0.142z^{-4}}{1+0.503z^{-1}+0.237z^{-2}} \\ \dfrac{0.228z^{-1}-5.508z^{-2}+0.609z^{-3}+1.072^{-4}}{1-0.276z^{-1}-0.106z^{-2}} \\ \dfrac{2.757z^{-1}-6.686z^{-2}+5.473z^{-3}-1.512z^{-4}}{1-1.633z^{-1}+0.690^{-2}} \end{bmatrix}^{\mathrm{T}}. \tag{6.9}$$

Shown in Figure 6.9, there are three turning periods detected in the initial phase: (i) from March 23 to October 25 in 2007, (ii) from June 2 to September 8 in 2008 and (iii) from December 1, 2008 to March 2, 2009. The second one is proven to be incorrect. Similarly, we re-estimate the internal model to track more recent market changes of the STI. The new OE model is given by

$$H_{\mathrm{oe,STI},2}(z) = \begin{bmatrix} \dfrac{-5.919z^{-1}-3.34z^{-2}+1.104z^{-3}-1.049z^{-4}}{1+0.0907z^{-1}-0.4484z^{-2}} \\ \dfrac{-0.4916z^{-1}-0.6866z^{-2}+7.735z^{-3}+2.524^{z}-4}{1-0.0783z^{-1}-0.3024z^{-2}} \\ \dfrac{2.447z^{-1}-1.774z^{-2}+6.728z^{-3}-7.004z^{-4}}{1-0.3793z^{-1}-0.0286^{-2}} \end{bmatrix}^{\mathrm{T}}. \tag{6.10}$$

A new turning period is identified, which is from June 22, 2011 to August 18, 2011. Structural breaks in the interest rate and the inflation rate of Singapore are also tested. The test results are given in Table 6.4 and Figure 6.9. The linkage between the internal model dynamics and the macroeconomic structural stability have once again been confirmed in the Singapore market. Both the interest rate and the inflation rate show their structural breaks in the first turning period. The second and third turning periods are also coincident with the structural break from either the interest rate or the inflation rate.

Table 6.3 Bai-Perron test results of structural breaks in the Hong Kong macroeconomy

Macroeconomic Series	Segment	Optimal Lag	Durbin-Watson stat.	B-G LM Test (p value)	UDmax	WDmax	$SupLR_T(l+1\|l)$			Break points
							$SupLR_T(2\|1)$	$SupLR_T(3\|2)$	$SupLR_T(4\|3)$	
Hong Kong interest rate	2005 – 2009	6	2.00	0.98	48.72*	60.60***	7.20	–	–	Oct., 2007
	2008 – 2012	3	1.95	0.74	34.75***	49.30***	19.93**	31.59***	6.36	Nov. 2008 Jun. 2009 Aug. 2011
Hong Kong inflation rate	2005 – 2009	6	1.98	0.96	31.94***	39.73***	48.16**	3.56	–	Aug. 2007 Jul. 2008
	2008 – 2012	4	1.99	0.90	22.77***	31.56***	22.86**	7.20	–	Sep. 2010 Jun. 2011

Note: * 10% level ** 5% level *** 1% level

Table 6.4 Bai-Perron test results of structural breaks in the Singapore macroeconomy

Macroeconomic Series	Segment	Optimal Lag	Durbin-Watson stat.	B-G LM Test (p value)	UDmax	WDmax	$\frac{SupLR_T(l+1\|l)}{SupLR_T(2\|1)}$	Break points
Singapore interest rate	2003 – 2009	4	2.01	0.27	33.99***	46.47***	7.16	Jul. 2007
	2008 – 2012	2	1.91	0.19	20.90***	21.19***	4.54	Jun. 2011
Singapore inflation rate	2003 – 2009	5	2.00	0.47	25.44***	31.01***	14.57	May 2007
	2008 – 2012	4	2.00	0.12	23.64***	23.64**	12.31	Apr. 2009

Note: * 10% level ** 5% level *** 1% level

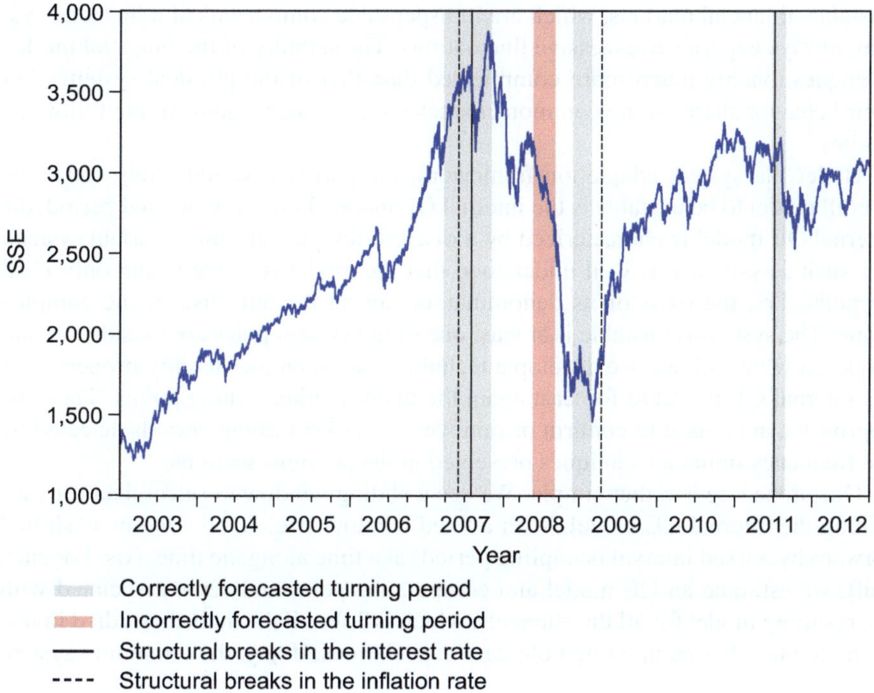

Fig. 6.9 Forecasted major market turning periods of the STI

6.3 A System Stability-Based Confirmation Method

Even though the frequency domain technique is capable of forecasting most of the major market reversals. It, however, does occasionally provide some false alarms. The results of the structural breaks in the macroeconomy cannot be used as the supplementary information to confirm the forecasts as it does not have the predictive ability. Inspired by the systems stability theory, we have discovered some very interesting connections between the unstable poles of the internal OE model and the major market turnings.

Stability is a very important property of a system as it indicates how the system evolves over time. It is an essential requirement for the design of a sensible control system since all the controlled physical and engineering systems have to be made stable. Although there is no standard definition, the financial stability is always considered as a situation that the financial system has the ability to facilitate and enhance economic processes, manage risks and absorb shocks [115]. In order to maintain the financial stability, similar to the engineering systems, some control actions will be taken into account such as adjusting monetary policies, adopting new stimulus packages and other regulations. A financial market is also a controlled system that it should be stable most of the time. When the financial system becomes unstable, financial markets, which are indispensable components of a financial system, always experience excessive fluctuations. The stability of the financial market is unquestionably much more complicated than that of the physical systems, but their behavior shares some common characteristics, which lead us to the follow-up studies.

Under the system adaptation framework, the part that would likely cause the overall model to be unstable is the internal OE model. For a certain fixed period, the internal OE model is characterized by a discrete-time linear time-invariant system. For such a system, it is well understood that the model is stable if and only if all its poles, i.e., the roots of its denominators, are on the unit disk of the complex plane. The system is unstable if at least one of the system poles are outside the unit circle. In what follows, we develop a technique based on the stability properties of the internal OE model to forecast again the major market turning points. This new approach can be used to confirm or remove the market turning periods detected by the frequency domain techniques presented in the previous sections.

Our procedure is rather simple. We use a sliding window to select data for estimating the internal OE model. With a fixed window size, ℓ, the window is shifted forward by a fixed interval (sampling period) at a time along the time axis. For each shift, we estimate an OE model and compute the poles and zeros associated with the resulting model for all the channels, which is 3 for all the markets studied in this monograph. After completing pole-zero cancelations, if any, the remaining system

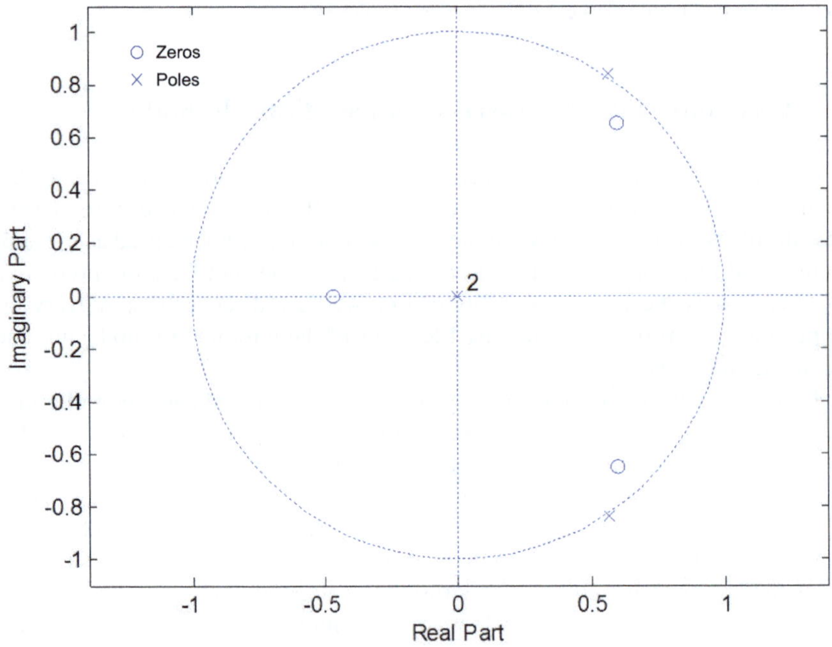

Fig. 6.10 Pole-zero map of an OE model channel

poles are then used to examine the model stability. For example, shown in Figure 6.10 is the pole-zero map of an OE model channel. It is clear that the resulting system is unstable as there are two poles that fall outside the unit circle.

As it is impractical to have a perfect pole-zero cancelation, a threshold value ρ_{pz} is introduced to determine if the cancelation of a pole and zero pair occurs or not. If the distance between the pole and the zero is less than ρ_{pz}, they are considered to be canceled. For an unstable system, at least one of the remaining poles is located outside the unit circle. For safety, we introduce another threshold $\rho_{\mathrm{p}} < 1$ to represent stability margin. Any pole with a magnitude larger than ρ_{p} is considered to be unstable. If any unstable pole is detected, we define the ending date of sliding window as an unstable point of the internal model. We have found that the unstable point of the internal OE model and the market turning periods are highly correlated to each other. As will be seen shortly, the unstable points of the internal OE model can serve as an effective means to confirm the major market turning periods detected by the frequency domain technique. With such a confirmation, the forecasting accuracy of the major market turnings can be greatly improved.

We proceed to test this system stability technique on the DJIA, the SSE, the HSI and the STI, respectively. According to the experiments that we have done on daily and weekly data, we find that a one-week interval is the best choice for the sampling period. Empirically, the sliding window size ℓ is set to 50, 60 or 75, which are approximately corresponding to one year to one and a half years. It means ℓ-weeks data prior to the starting point are used to estimate the OE model. It should be noted that this is a purely empirical selection. Also, to be more practical, we set $0.95 \leq \rho_{\mathrm{p}} < 1$, and $0.01 \leq \rho_{\mathrm{pz}} \leq 0.1$.

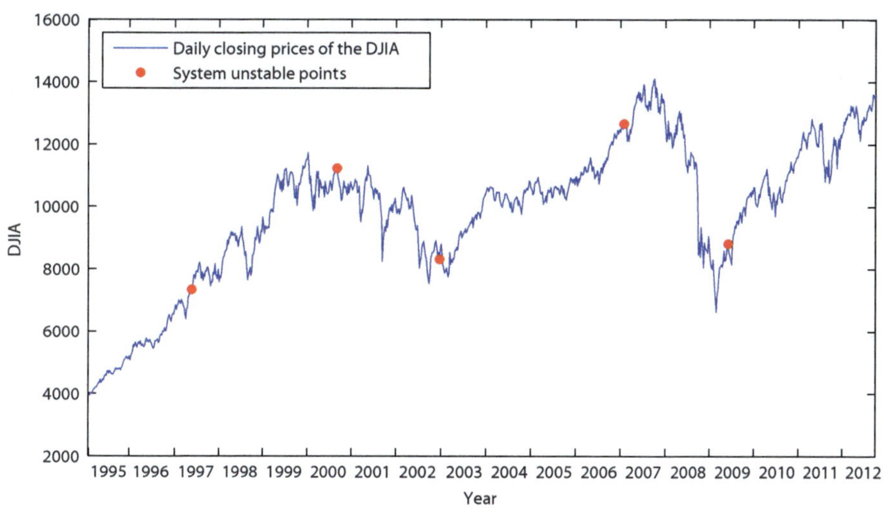

Fig. 6.11 Unstable points on Channel 2 of the DJIA with $\ell = 50$, $\rho_{\mathrm{p}} = 0.97$, $\rho_{\mathrm{pz}} = 0.1$

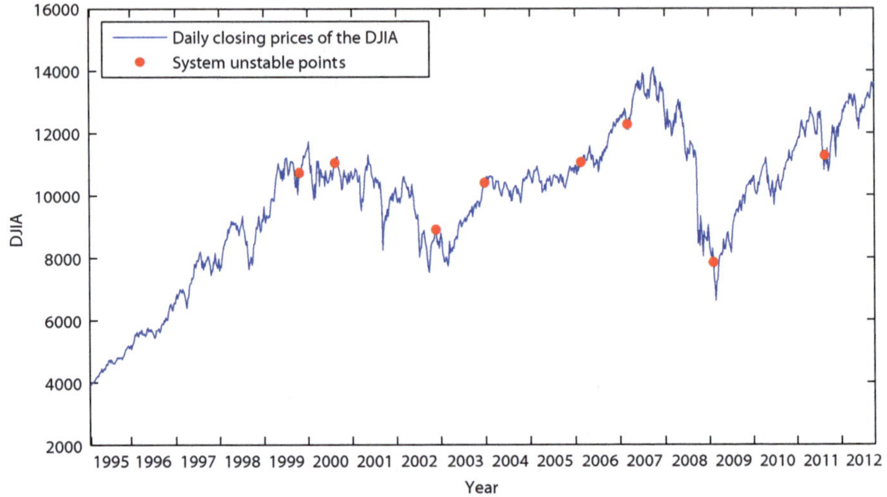

Fig. 6.12 Unstable points on Channel 2 of the DJIA with $\ell = 60$, $\rho_p = 0.99$, $\rho_{pz} = 0.02$

Fig. 6.13 Unstable points on Channel 2 of the DJIA with $\ell = 75$, $\rho_p = 0.96$, $\rho_{pz} = 0.05$

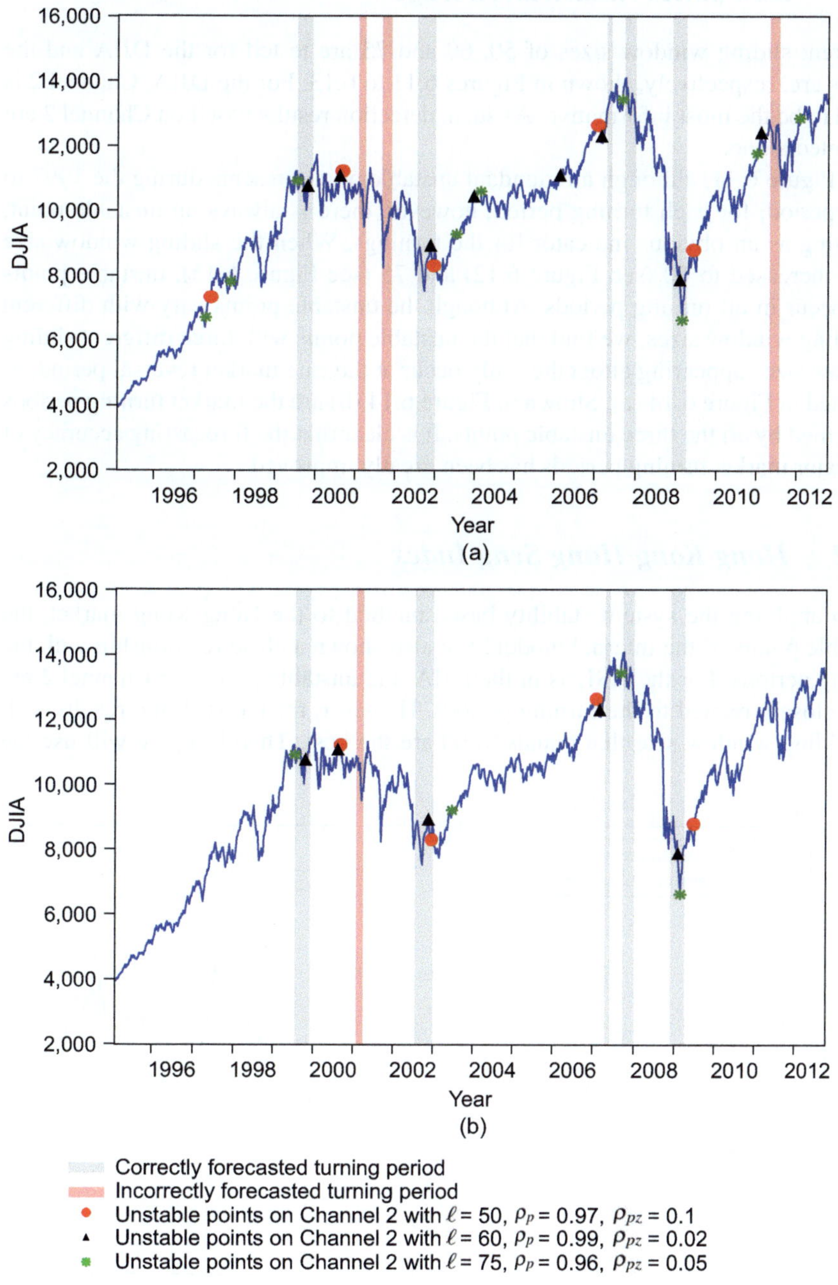

Fig. 6.14 Turning periods of the DJIA confirmed by the internal model unstable points

6.3.1 Dow Jones Industrial Average

Different sliding window sizes of 50, 60 and 75 are tested for the DJIA and the results are, respectively, shown in Figures 6.11 to 6.13. For the DJIA, Channel 2 is found to be the most informative. As such, detection results based on Channel 2 are presented alone.

In Figure 6.11, although a redundant unstable point presents during the 1997 to 1998 period, for each turning period, however, there is always an unstable point, behaving as an obvious indicator for the turnings. When the sliding window size (ℓ) is increased to 60 (see Figure 6.12) and 75 (see Figure 6.13), unstable points also occur in all turning periods. Although the unstable points vary with different sampling window sizes, we find that the unstable points with three different sliding window sizes appearing altogether only occur at the true market reversal period, as depicted in Figure 6.14 (a). Shown in Figure 6.14 (b) are the market turning periods confirmed by all the three unstable points. It is clear that the forecasting accuracy of the major market turning periods has been greatly improved.

6.3.2 Hong Kong Hang Seng Index

When applying the system stability-based method to the Hong Kong market, the unstable points of the internal model have also shown a close relationship with the turning periods. For the HSI, as in the DJIA, the unstable points on Channel 2 are most closely related to the turning periods. However, for the HSI, the results with the sliding window size that equals to 60 are the best. Therefore, we will use the

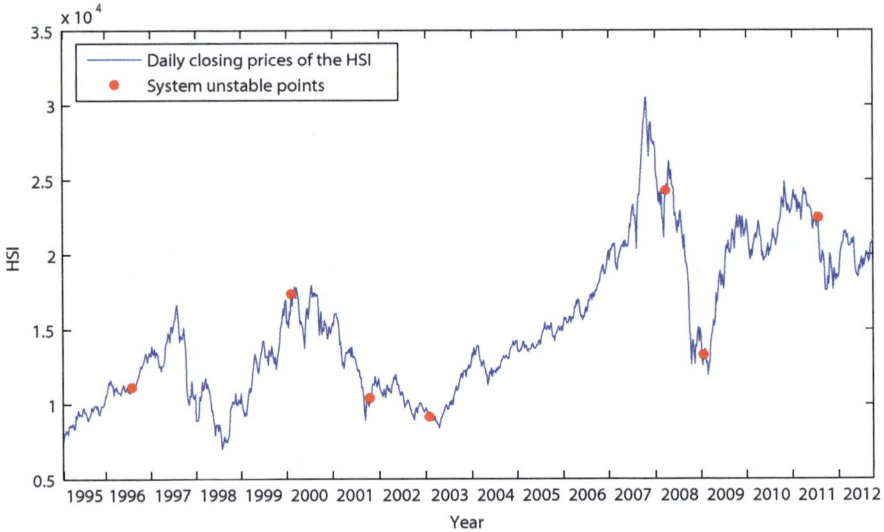

Fig. 6.15 Unstable points on Channel 2 of the HSI with $\ell = 60$, $\rho_p = 0.97$, $\rho_{pz} = 0.05$

Fig. 6.16 Turning periods of the HSI confirmed by the internal model unstable points

unstable points with $\ell = 60$ (see Figure 6.15) for the confirmation of the market reversals.

In Figure 6.15, the unstable points of the internal model appear in every turning period after the rise in 1998. Similarly, as it is done for the DJIA, we integrate the turning periods forecasted by the frequency domain technique with the unstable points (see Figure 6.16) to remove incorrect forecast.

6.3.3 Singapore Straits Times Index

Different from the DJIA and the HSI, the unstable points on Channel 1 of the internal model are most significantly related to the major market turnings of the STI. As shown in Figure 6.17, setting the sliding window size to 75, there are unstable points in almost every turning period. By integrating the turning periods forecasted by the frequency domain technique together with the internal model unstable points (see Figure 6.18), we can completely remove the false detection.

6.3.4 Shanghai Stock Exchange Composite Index

The detection of the unstable points of the internal model in the China market is more complicated. If the previous detection criteria are used, the unstable points obtained are not informative at all in confirming the turning periods for the SSE. We thus slightly loosen the criteria for the SSE in order to get more meaningful confirmation signals. We set $0.01 \leq \rho_{\mathrm{pz}} \leq 0.1$ and $0.91 \leq \rho_{\mathrm{p}} < 1$.

Fig. 6.17 Unstable points on Channel 1 of the STI with $\ell = 75$, $\rho_{\mathrm{p}} = 0.97$, $\rho_{\mathrm{pz}} = 0.05$

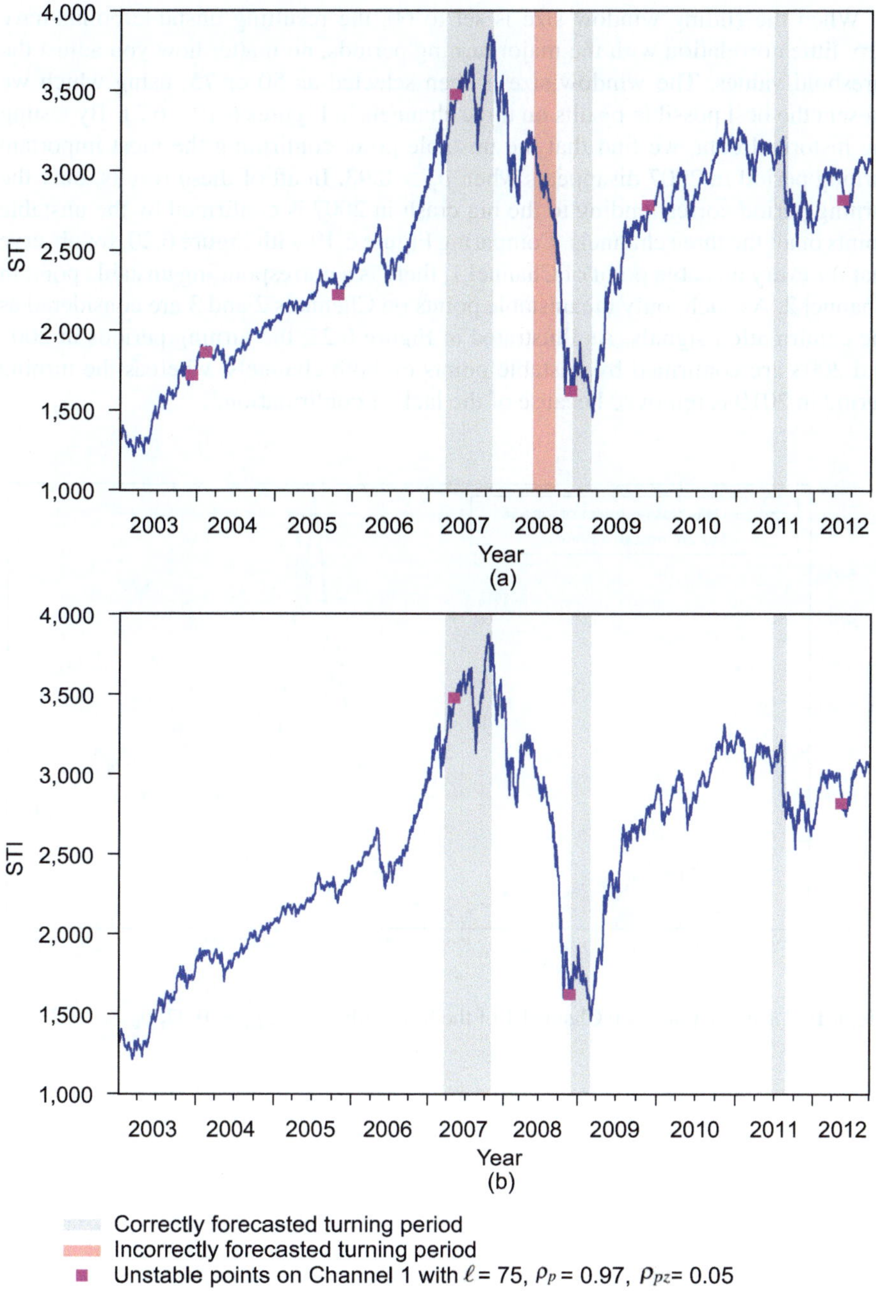

Fig. 6.18 Turning periods of the STI confirmed by the internal model unstable points

When the sliding window size is set to 60, the resulting unstable points have very little correlation with the major turning periods, no matter how you adjust the threshold values. The window size is then selected as 50 or 75, using which we present the best possible results on three channels in Figures 6.19 – 6.21. By testing the historical data, we find that the unstable point confirming the most important turning period in 2007 disappears when $\rho_p > 0.93$. In all of these results, only the turning period corresponding to the big crash in 2007 is confirmed by the unstable points of all the three channels. Comparing Figure 6.19 with Figure 6.20, we observe that for every unstable point of Channel 1, there is a corresponding unstable point in Channel 2. As such, only the unstable points on Channels 2 and 3 are considered as the confirmation signals. As illustrated in Figure 6.22, the turning periods in 2007 and 2008 are confirmed by unstable points on both channels, whereas the turning period in 2010 is removed because of the lack of confirmation.

Fig. 6.19 Unstable points on Channel 1 of the SSE with $\ell = 50$, $\rho_p = 0.97$, $\rho_{pz} = 0.02$

Fig. 6.20 Unstable points on Channel 2 of the SSE with $\ell = 50$, $\rho_p = 0.93$, $\rho_{pz} = 0.1$

Fig. 6.21 Unstable points on Channel 3 of the SSE with $\ell = 75$, $\rho_p = 0.95$, $\rho_{pz} = 0.01$

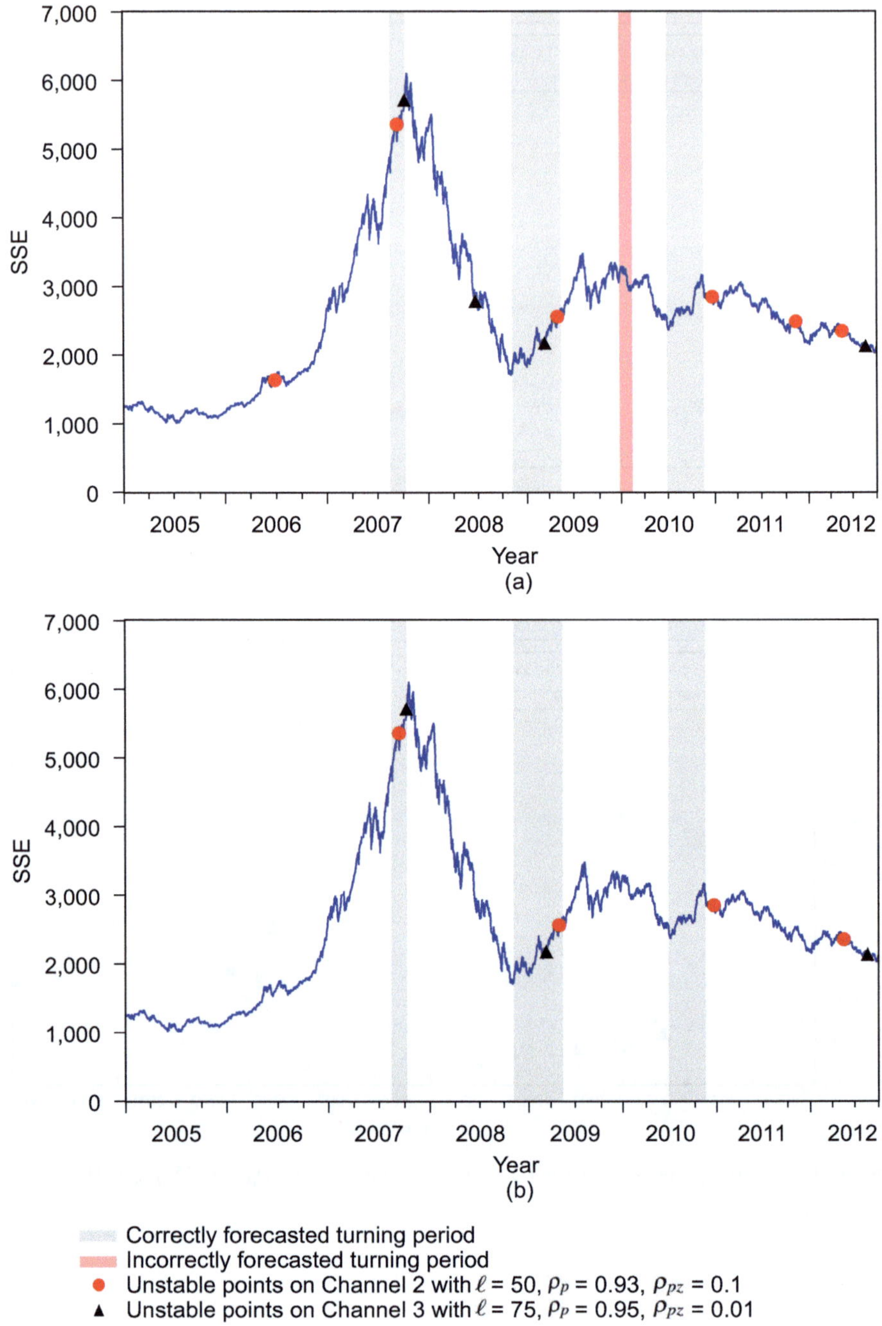

Fig. 6.22 Turning periods of the SSE confirmed by the internal model unstable points

Chapter 7
Technical Analysis Toolkit

7.1 Introduction

We present in this chapter some basic features and functionalities of a toolkit for technical analysis of stocks (T-TAS) developed under a MATLAB environment (Version 6.5 and above). The initial version of the toolkit [34] was a result of a final year project conducted in the Department of Electrical and Computer Engineering, National University of Singapore. Since then, the toolkit has been significantly revised and enhanced to incorporate new features, such as those related to the system adaptation framework documented in this monograph. Shown in Figure 7.1 is the main interface of the T-TAS.

The toolkit is developed for the purpose of providing an easy-to-use yet powerful platform to analyze the stock markets or financial markets in general. There are two main features of the toolkit. One is a user-friendly graphical interface, implemented by the MATLAB graphical user interface (GUI) tools, to provide an intuitive and interactive environment. From online data loading to data analysis, it is just a single click away. It also provides a clear display of the analyzed results including the stock prices and volumes, plotting of indicators and trading signals. The other feature is its advanced functionalities. The T-TAS is linked to the historical as well as real-time prices. As such, the data analysis could be performed at weekly, daily and even intraday frequencies. It provides basic and advanced technical analysis as well as some newly developed functions related to the system adaptation framework. These functionalities can be used in a simplest form that the user could use them without in-depth programming or chart reading skills. This chapter is aimed to serve as a user manual for the toolkit. However, in order to appreciate all functions implemented, some basic knowledge of the technical analysis would certainly be helpful.

The T-TAS has been fully tested for counters listed on the NYSE, NASDAQ, Singapore Stock Exchange, Hong Kong Stock Exchange, Shanghai and Shenzhen Stock Exchanges. The toolkit should work for the markets in Australia, Indonesia, India and almost all markets worldwide. It can also be extended to the analysis of other financial markets. Even though there are many financial analysis (generally expensive) software platforms available in the market, the T-TAS offers a cheap

X. Zheng & B.M. Chen: *Stock Market Modeling and Forecasting*, LNCIS 442, pp. 123–147.
DOI: 10.1007/978-1-4471-5155-5_7

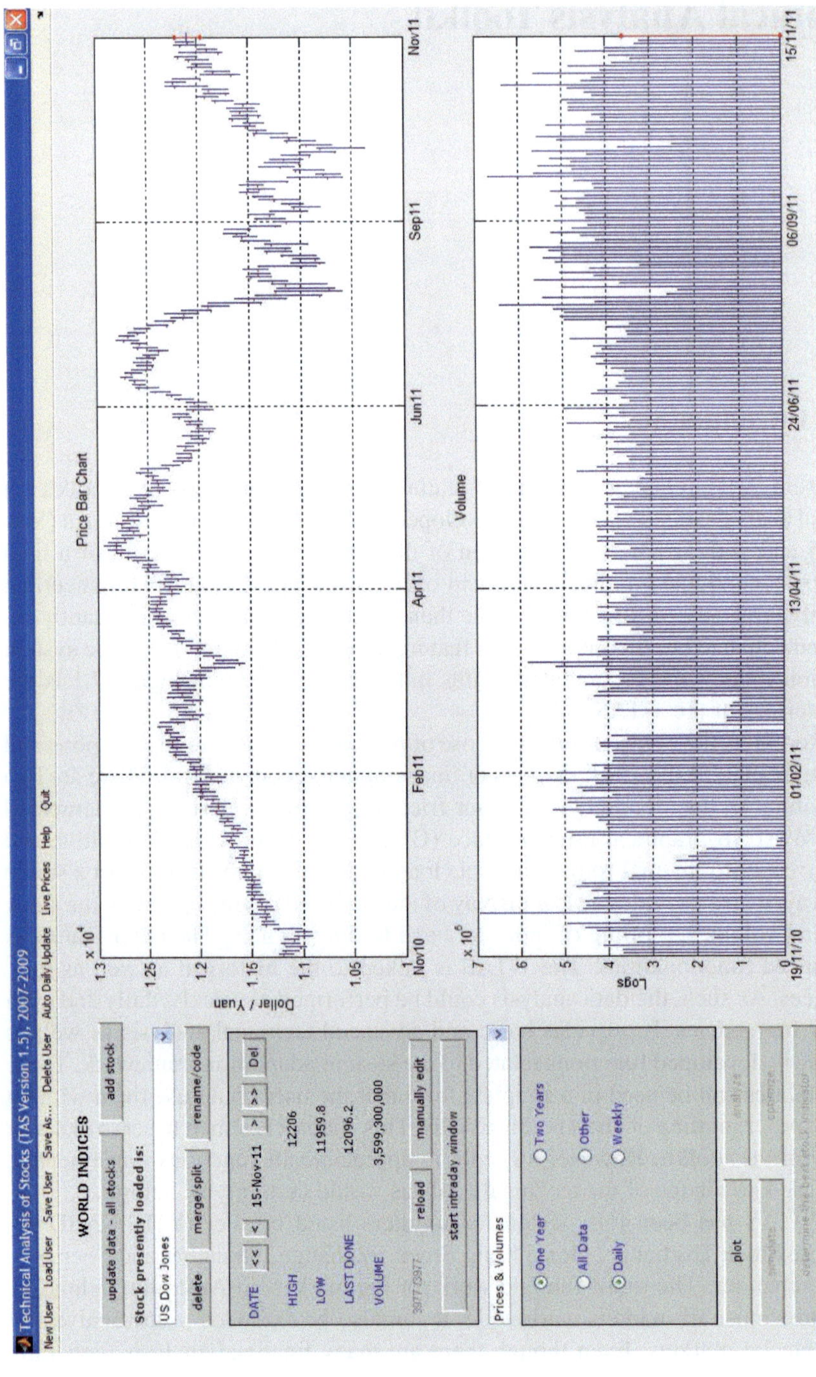

Fig. 7.1 Main panel of the T-TAS

alternative for personal trading and for scientific research. It can be freely modified to accommodate more and/or few functions.

Interested readers are referred to a web page maintained by the authors, which is hosted at http://uav.ece.nus.edu.sg/~bmchen/, for the most up-to-date information on the toolkit.

7.2 T-TAS Functions

In this section, we highlight some key functions provided by the T-TAS, which include user management, data manipulation, auto data loading system, technical analysis and trading performance analysis, as well as technical tools associated with the system adaptation framework.

7.2.1 User and Data Management

The T-TAS groups and manages stock counters by user accounts. Within a particular user account, one can manipulate a list of stock counters of interest by using the adding and deleting functions. Shown in Figure 7.2 is a list of counters maintained by the authors in the toolkit under a user name 'World Indices'.

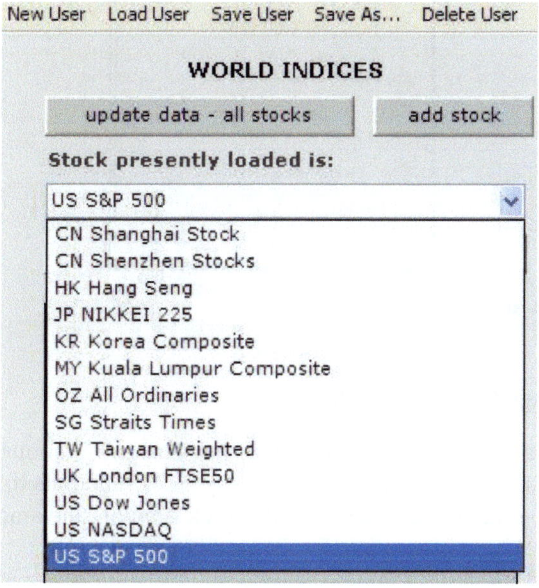

Fig. 7.2 A typical user profile

7.2.1.1 Create New Users

To start using the toolkit, one first needs to use the function 'New User' on the main panel to create a user account. After the account being created, a user can then proceed to add in stock counters of his/her choice from the stock markets worldwide, so long as their are captured by the Yahoo Finance [131]. The toolkit also allows its users to add more accounts and delete some unwanted.

7.2.1.2 Add and Delete Stock Counters

Adding and deleting stock counters are rather easy in the T-TAS. Within a user account, say for example the 'World Indices' in Figures 7.1 and 7.2, one can add in a new counter by clicking on the 'add stock' function. A new window (see Figure 7.3) will pop out prompting the user to enter the symbol and name of a stock counter that one wants to add in the watch list. As all stock data are to be downloaded from the Yahoo Finance [131], the stock symbol entered has to be consistent with that adopted by Yahoo. The counter name, however, can be any free text. When a new counter is added, its historical data will automatically be downloaded online from [131]. The 'delete' function under a specific counter (see Figure 7.1) can be used to delete an unwanted counter.

Fig. 7.3 Add a new stock counter

7.2.1.3 Manage Historical Data

To adopt changes in the market, the T-TAS has been programmed to allow users to manually edit the stock data. The toolkit provides functions to merge or split a counter shares, to rename a counter, modify stock symbols and amend missing data.

1. *Merge or Split Stock Shares*
   Companies may merge or split their stock shares for many reasons, which would cause a big jump in the price series. With the function of 'Merger/Split', users can set the ratio of the share merger or split (see Figure 7.4) to adjust accordingly all the historical data captured in the database.

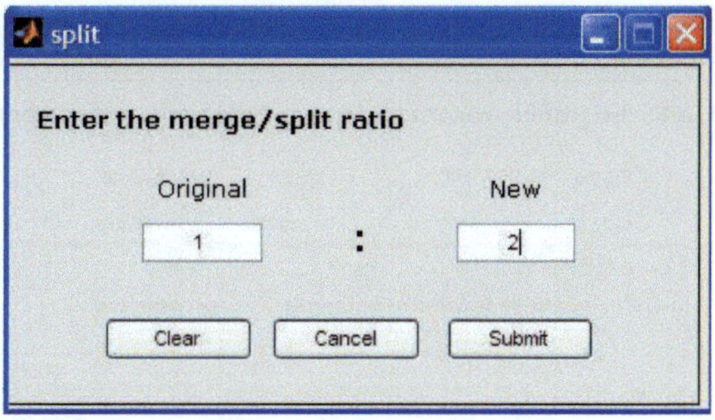

Fig. 7.4 Merger or split of stock shares

2. *Change Stock Counter's Name and Code*
   The toolkit provides an easy way to change the name and symbol of a specific counter. A pop-out window associated with the 'rename/code' function is shown in Figure 7.5, which can be used to modify the new name and code of a particular stock counter.

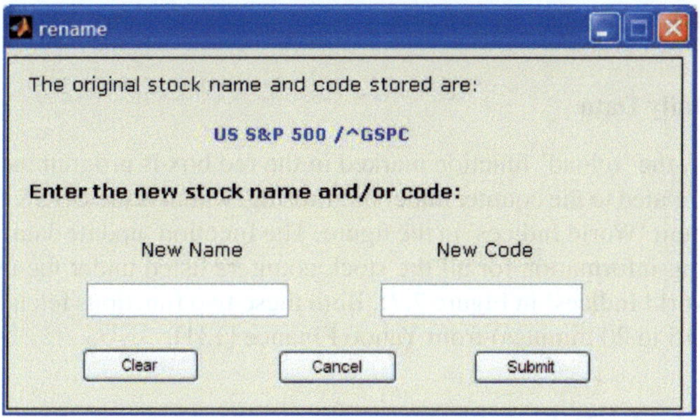

Fig. 7.5 Change of counter's name and/or code

3. *Edit Stock Prices and Volumes*
   The toolkit allows manual editing of stock prices and volume with the 'manually edit' function. Figure 7.6 is a pop-out window associated with this function, which allows the user to amend the wrongly captured historical data.

7.2.2 Online Data Loading System

The T-TAS is programmed to download all historical data and new daily data online from Yahoo Finance [131]. For intraday real-time prices, the toolkit is instructed to fetch online data provided by Google Finance [55] instead. The online data loading

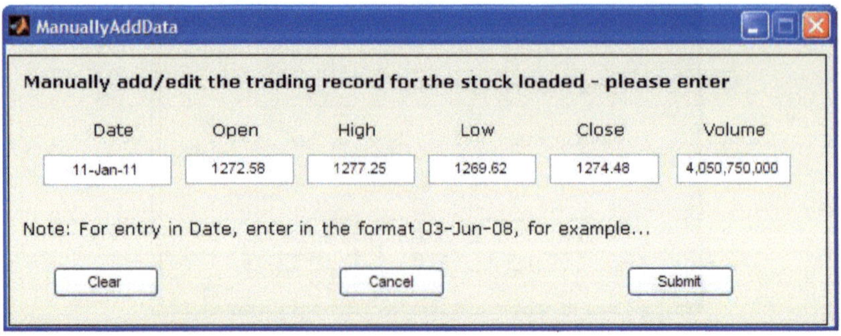

Fig. 7.6 Stock data editing

system of the T-TAS is comprehensive. It allows users to update share information for either a particularly selected stock counter or a group of the counters under a specific user account. It can also be set to automatically download fresh data from the online systems at a specific time, say, for example, 30 minutes after the market is closed. As highlighted in Figure 7.6, the stock data captured include the daily high, low, open, close prices and the volume.

7.2.2.1 Daily Data

In Figure 7.7, the 'reload' function marked in the red box is programmed to update information related to the counter under monitoring, which is the US S&P 500 inside the user account 'World Indices' in the figure. The function 'update data – all stocks' is for updating information for all the stock counters listed under the user account (which is 'World Indices' in Figure 7.7). Both these two functions fetch market data (delayed by 15 to 20 minutes) from Yahoo Finance [131].

7.2.2.2 Intraday Data

Intraday analysis is based on more frequently updated 'real-time' prices from the Google Finance [55]. Once the 'start intraday window' function (see Figures 7.1 and 7.7) is activated, a new interface panel (see Figure 7.8) will pop out to automatically download and display the price and volume of the stock counter under investigation at a pre-selected frequency. Users can analyze the intraday data using the technical tools provided. This function is particularly useful for active traders.

7.2.2.3 Automatic Daily Update

The 'auto daily update' function (see Figure 7.9) is programmed for inactive traders, who do not bother to monitor the market very frequently but want to keep their records up to date, particularly for counters traded in an increment of 0.1 cent in the Hong Kong and Singapore markets. The historical data captured by the Yahoo

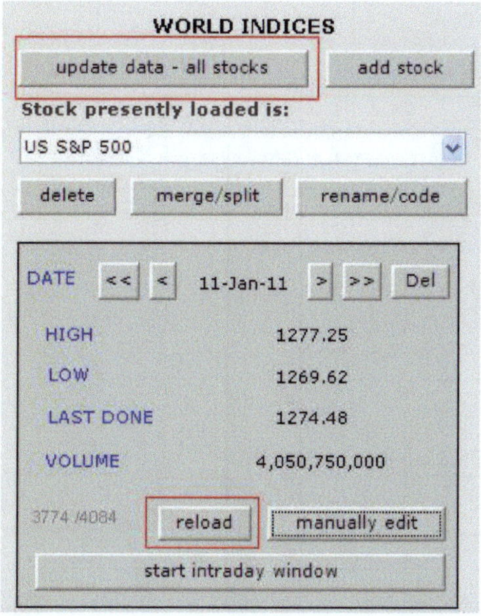

Fig. 7.7 Stock data update

Finance [131] is only accurate up to 2 decimal digits. Everything below one cent is either rounded up or rounded down. As such, the historical data provided the Yahoo Finance [131] for a larger amount of stock counters in the Singapore and Hong Kong markets are useless. Such a problem can be resolved by the 'auto daily update' function, which can correctly download the fresh (instead of historical) data from the Yahoo Finance online system [131].

7.2.3 *Technical Analysis*

Technical analysis is the core of the T-TAS. The toolkit provides many widely used technical indicators and trading rules. With these indicators and trading rules, users can analyze a specific stock, optimize parameters associated with the indicators, simulate and compare investment performance. Even though we find the toolkit and rules to be useful, we would like to emphasize that we bear no responsibilities whatsoever to any gain or loss that one might generate from the market.

1. *Technical Analysis Indicators*
   The following technical indicators are programmed in the T-TAS: (i) Moving Average Convergence/Divergence (MACD) (line and histogram indicators), (ii) Stochastic Oscillator (fast and slow stochastic indicators), (iii) Relative Strength Index (RSI), (iv) Elder-Ray Indicator [43], (v) Ben's Price-Volume Indicator (a self-defined indicator by the second author, which takes trading volume into consideration), (vi) Candle Stick Chart, and lastly, (vii) Bollinger Band. All these

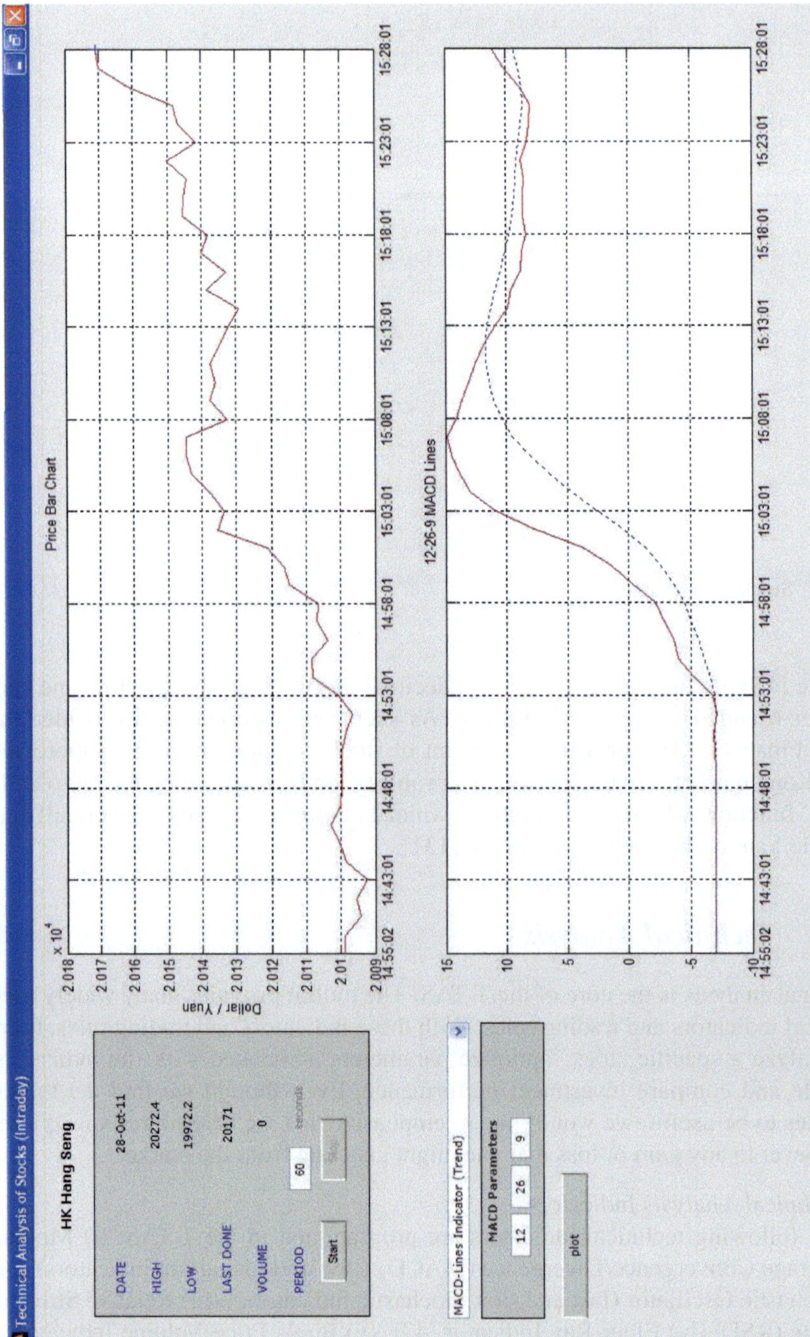

Fig. 7.8 Intraday analysis panel

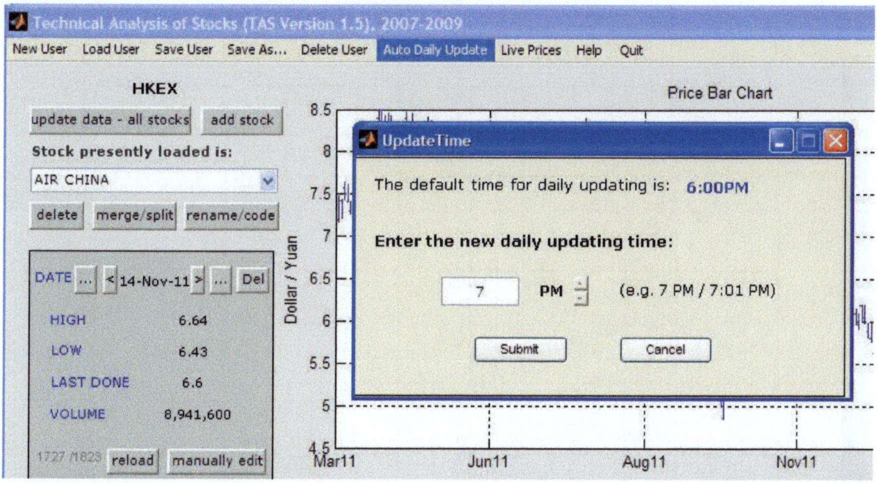

Fig. 7.9 Automatic daily update

are illustrated in Figures 7.10 to 7.18, respectively. The parameters used in these technical indicators are adjusted and data range can be freely selected, providing a great flexibility for the analysis.

2. *Trading Rules*
   The T-TAS includes a set of trading rules, based on which one can carry out necessary simulation, optimization and other analysis. Table 7.1 lists all the trading rules implemented in the toolkit. For the Japanese candlesticks, its patterns and trading signals could be found at http://www.candlesticker.com/.

3. *Indicator Evaluation*
   Based on the rules given in Table 7.1, the toolkit users can evaluate the effectiveness of a technical indicator on a particular stock counter by using the 'analyze' function on the main panel (see Figure 7.1). Figure 7.19 shows a typical simulation result conducted for a stock counter with the MACD indicator. The analysis result is summarized in a pop-up window.

4. *Indicator Parameter Optimization*
   For a technical indicator and trading rules adopted, the 'optimize' function on the main panel (see Figure 7.1) is to determine an optimal parameter setting for the indicator, which would yield the best performance, i.e., the maximum investment return over the period selected.

5. *Investment Simulation*
   The 'simulate' function on the main panel (see Figure 7.1) is used to simulate the performance of an investment based on a selected indicator and its associated rules over the period of interest. The function will return a total number of trades and the profit (or loss) over the period.

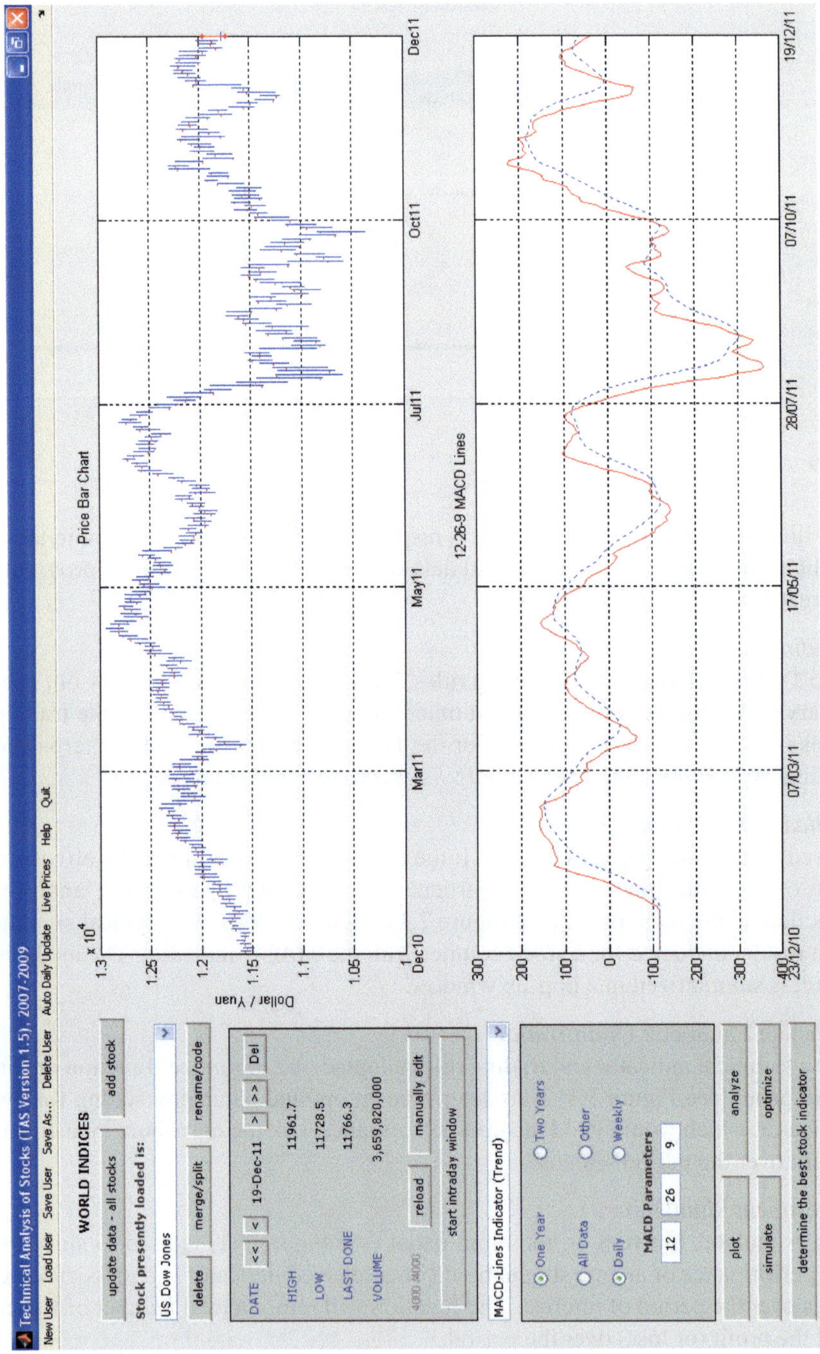

Fig. 7.10 MACD line indicator

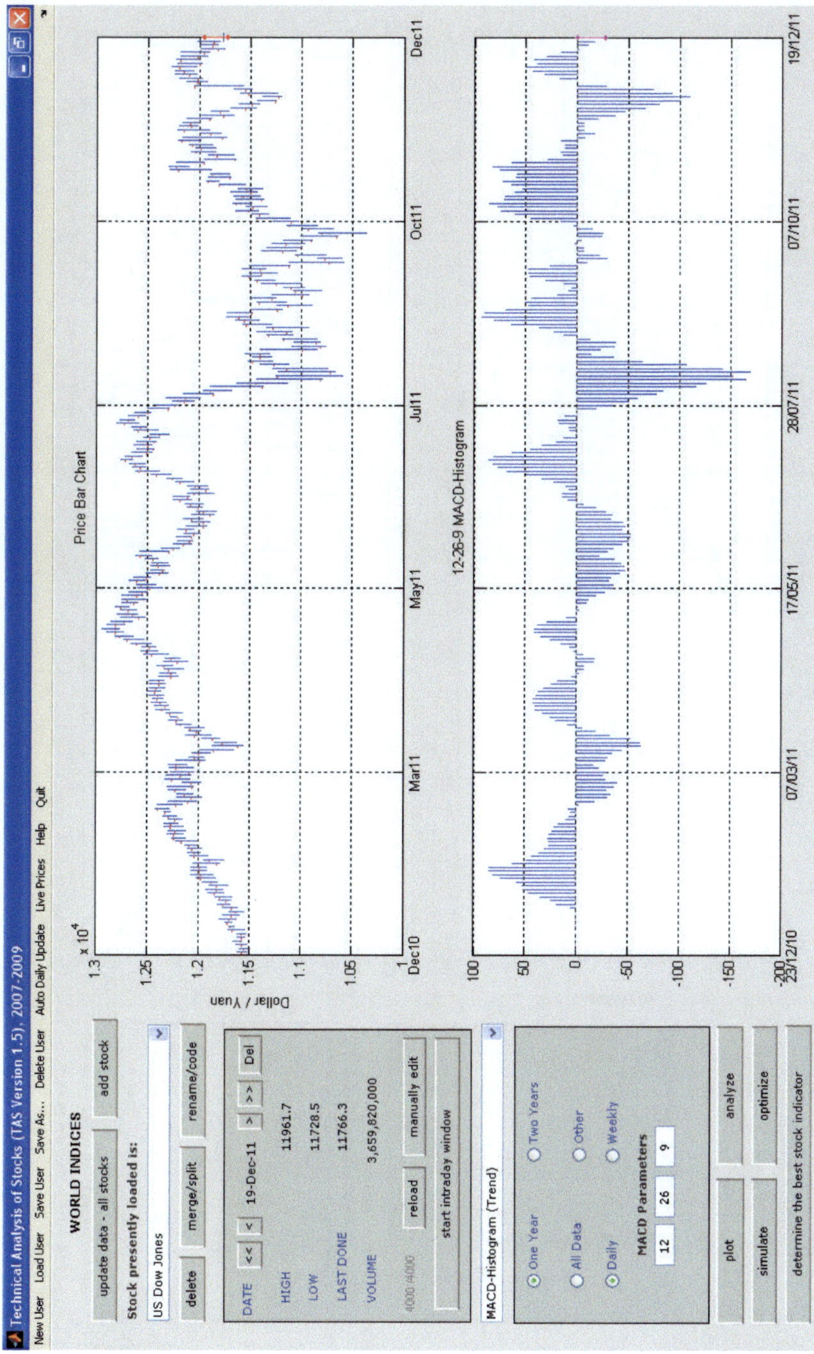

Fig. 7.11 MACD histogram indicator

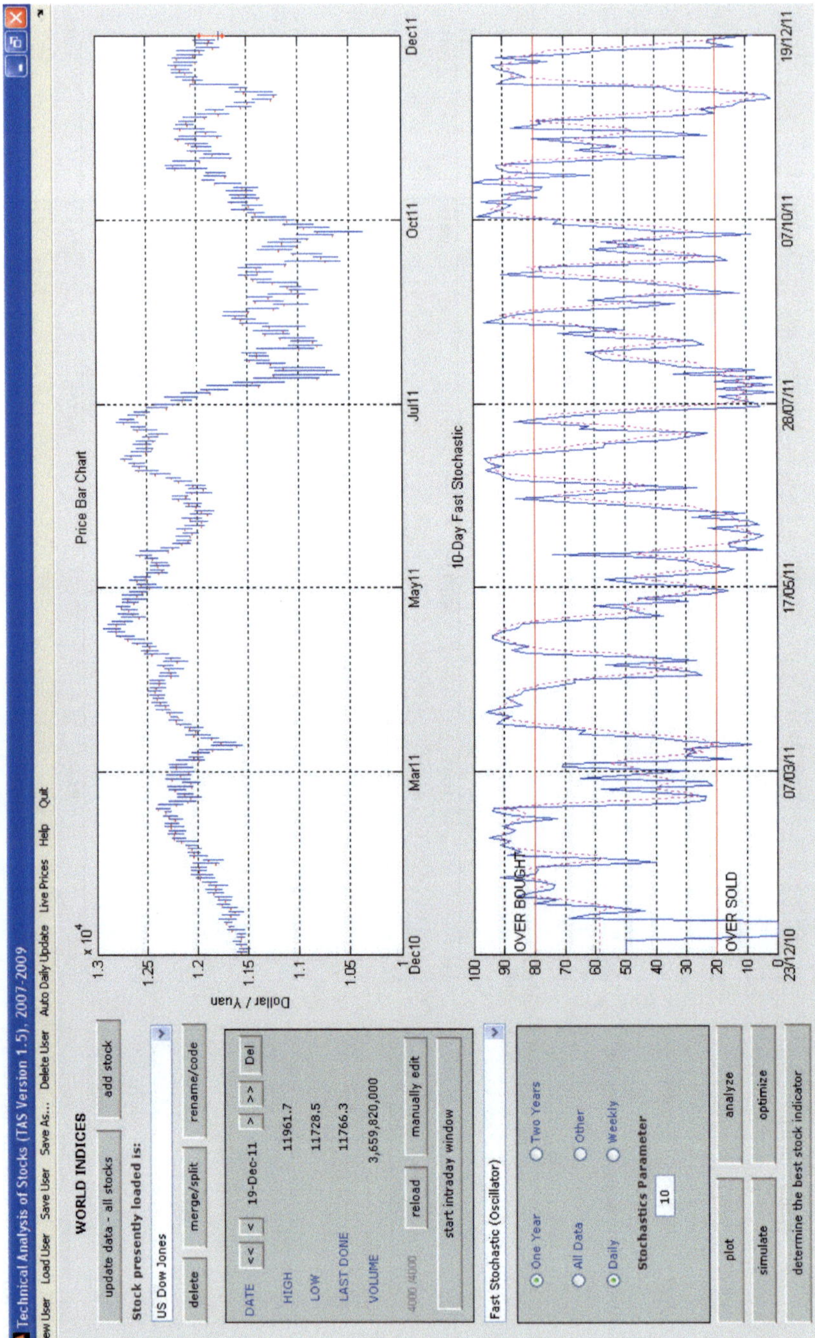

Fig. 7.12 Fast stochastic indicator

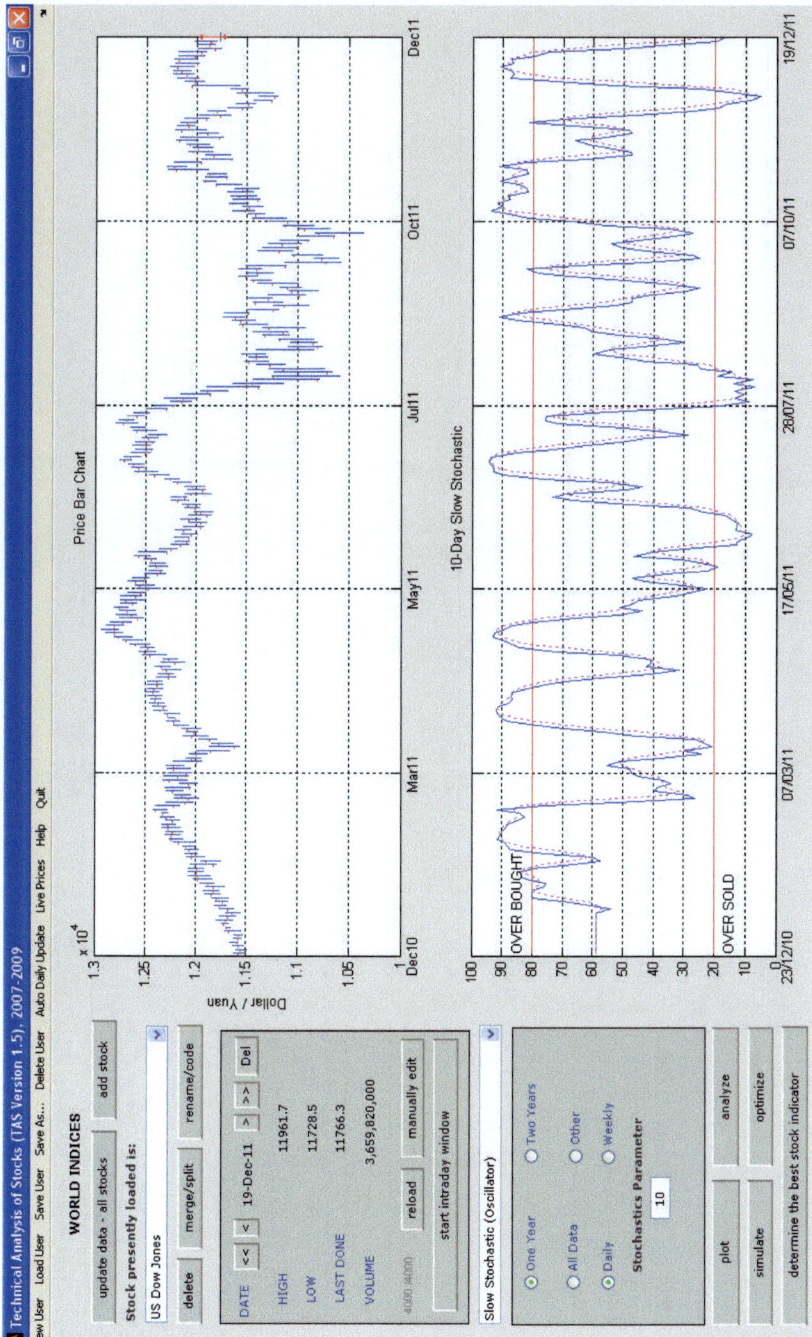

Fig. 7.13 Slow stochastic indicator

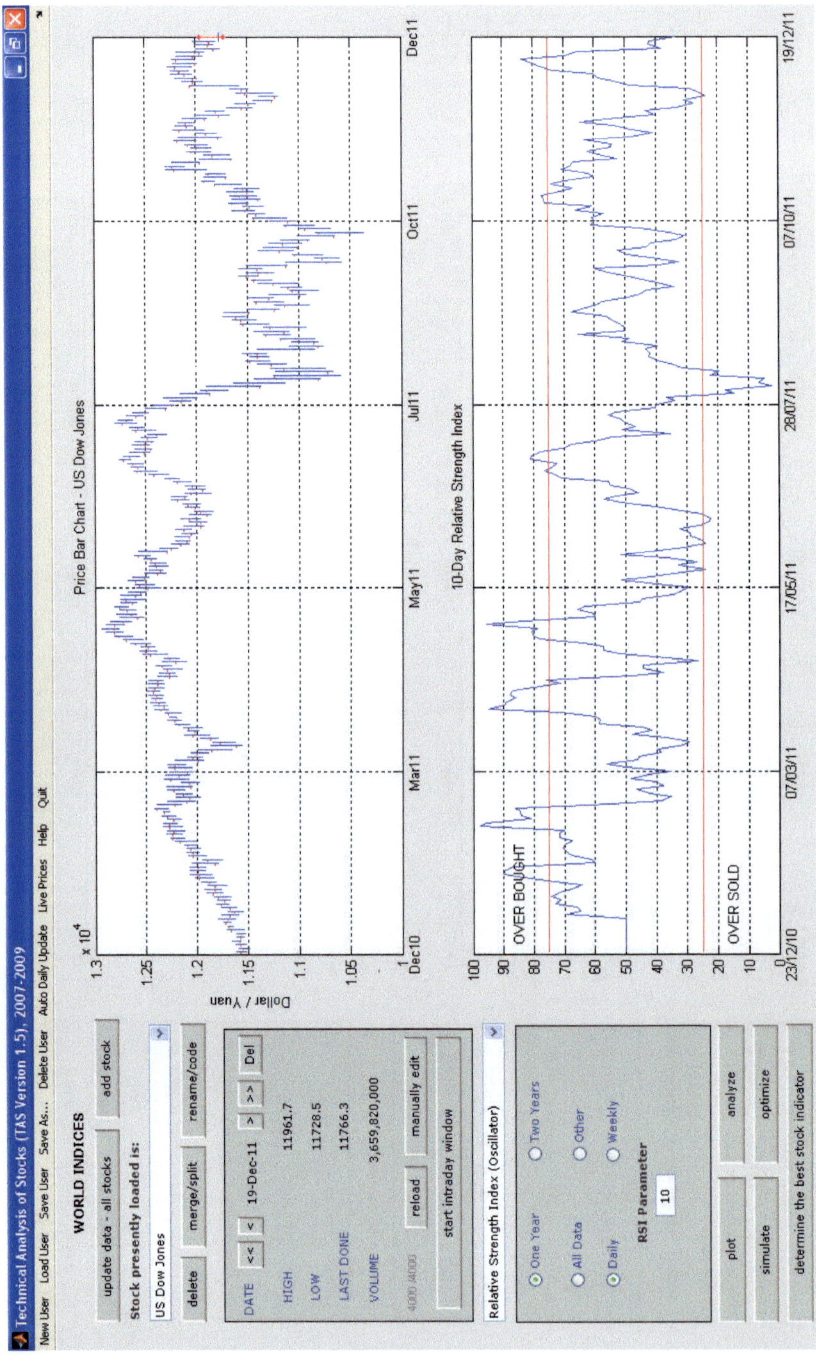

Fig. 7.14 Relative strength index

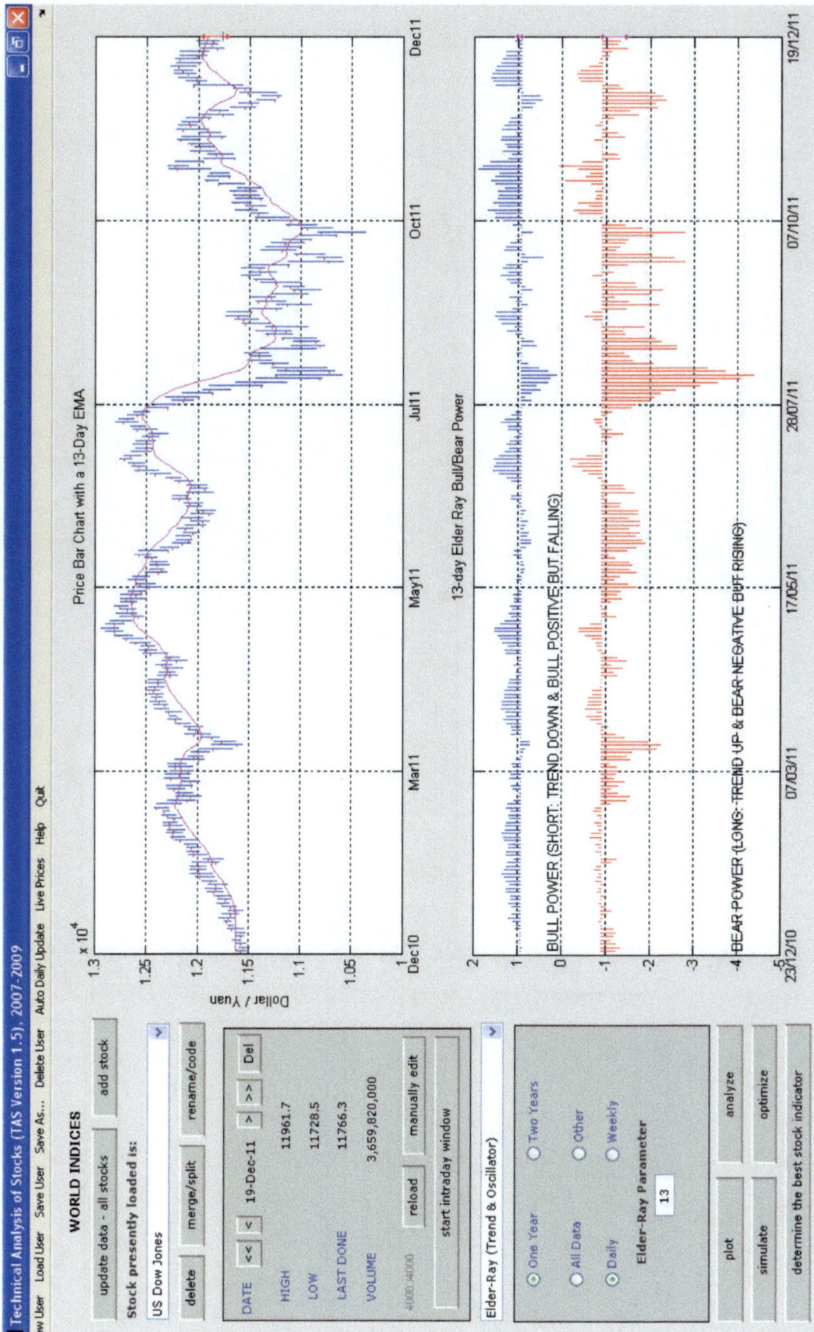

Fig. 7.15 Elder-Ray indicator

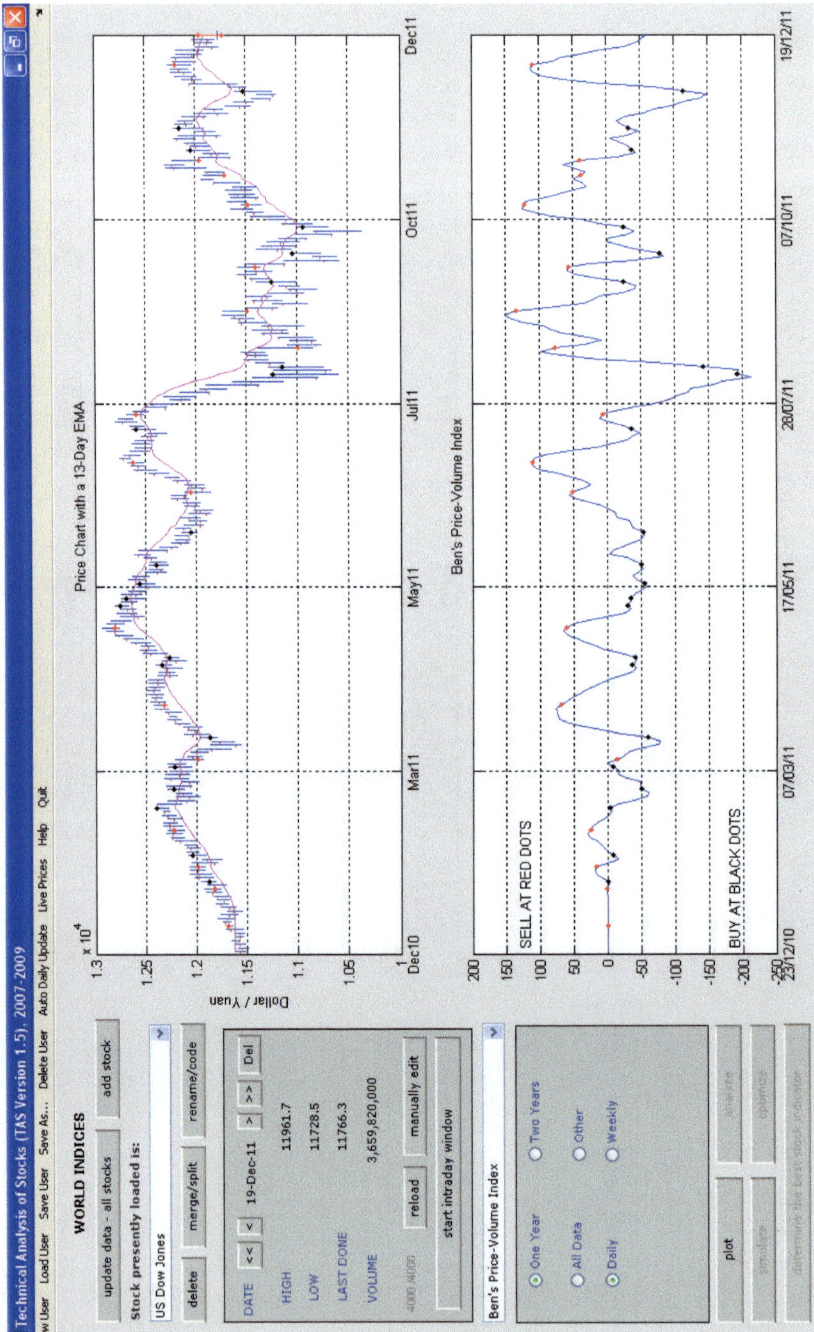

Fig. 7.16 Ben's Price-Volume indicator

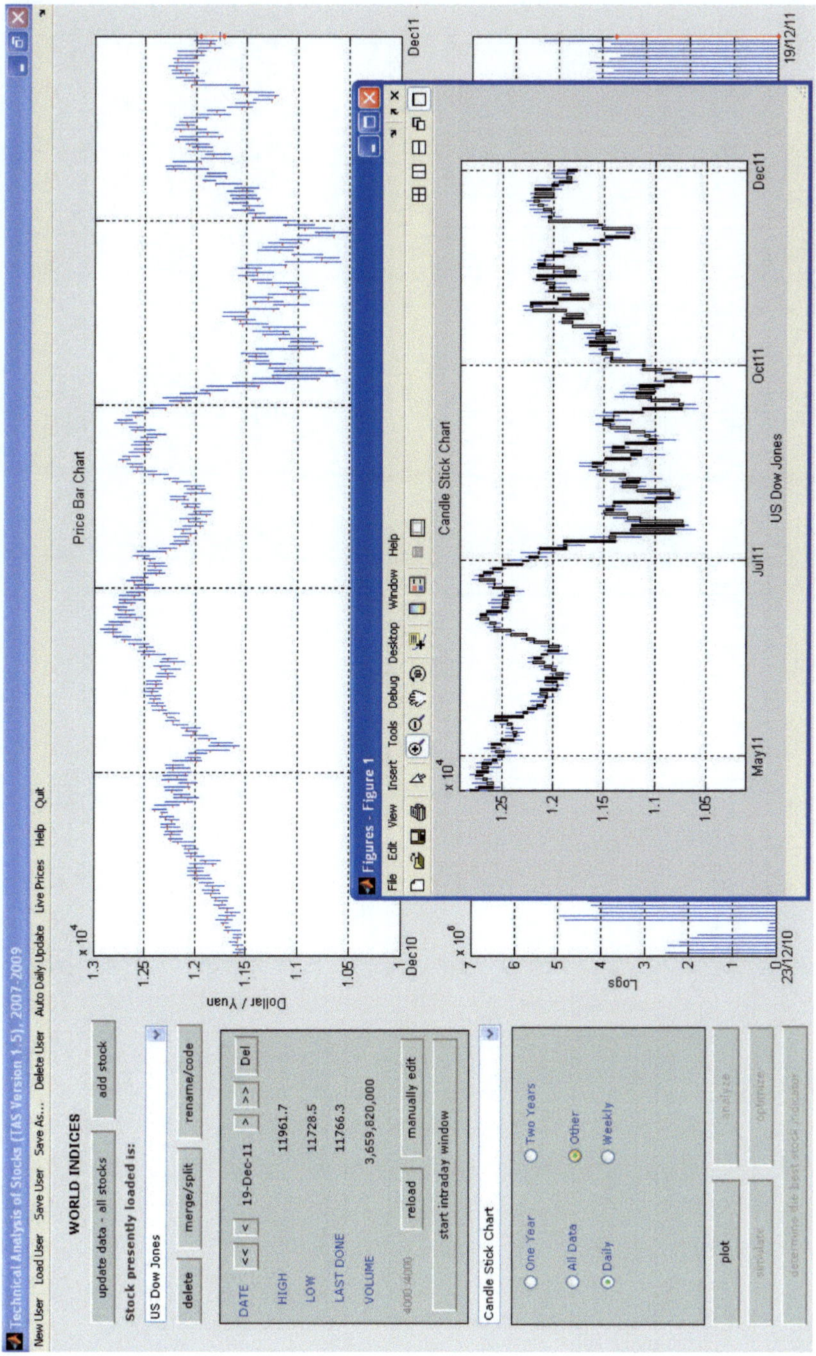

Fig. 7.17 Candle stick chart

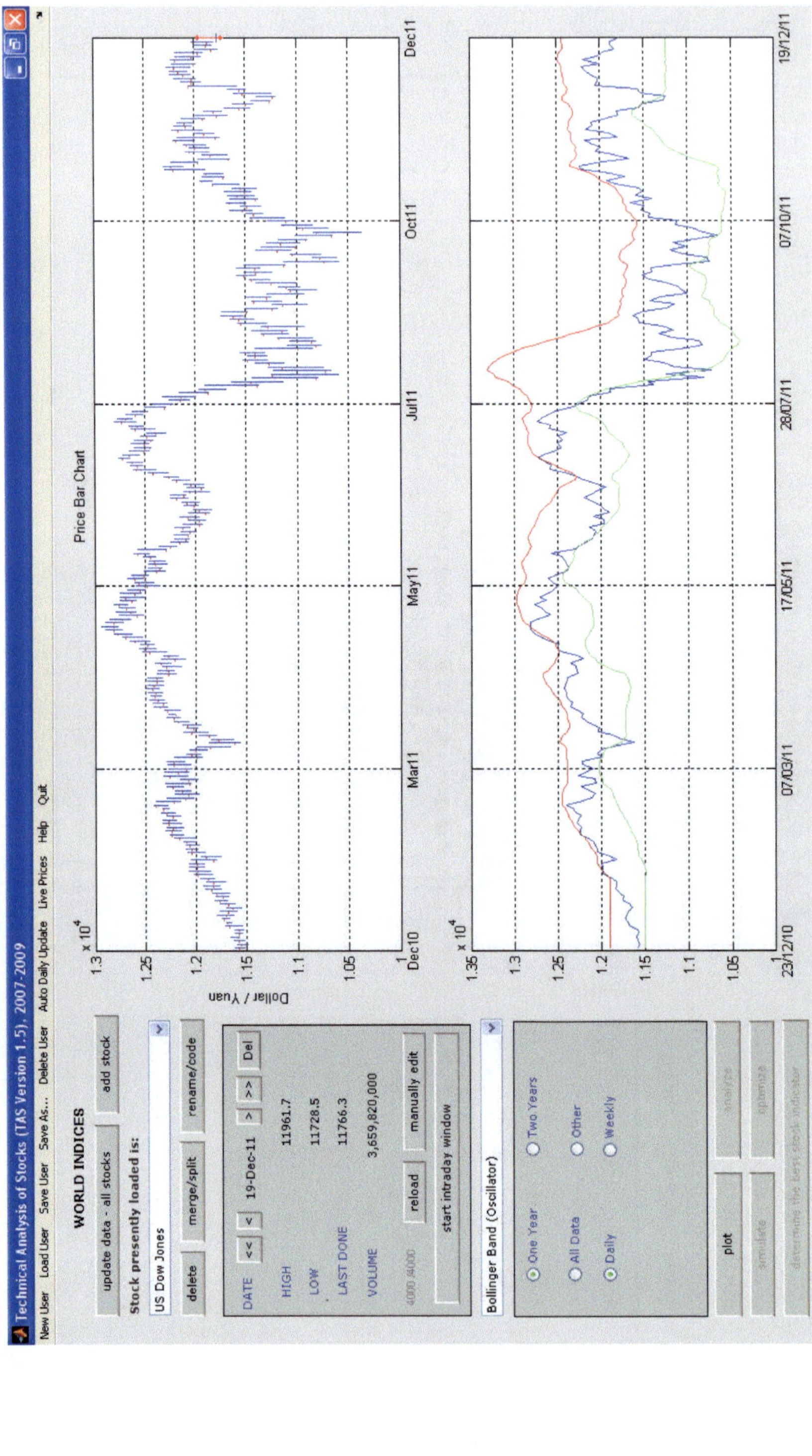

Fig. 7.18 Bollinger band indicator

Table 7.1 Trading rules adopted in T-TAS

TRADING RULES	BULL	BEAR
MACD Moving Average crossover	Go Long when MACD line > Signal line Cover Long when MACD line < Signal line	Go Short when MACD line < Signal line Cover Short when MACD line > Signal line
MACD centerline crossover	Go Long when MACD Line > 0 Cover Short when MACD Line < 0	Go Short when MACD Line > 0 Cover Long when MACD Line < 0
MACD line and histogram combination	Go Long when MACD Line > 0 and MACD Histogram > 0 Cover Long when MACD Line < 0 or MACD Histogram < 0	Go Short when MACD Line < 0 & MACD Histogram < 0 Cover Short when MACD Line > 0 or MACD Histogram > 0
MACD conservative histogram	Go Long when MACD Histogram > 0 Cover Long when MACD Histogram makes first peak	Go Short when MACD Histogram < 0 Cover Short when MACD Histogram makes first trough
RSI Centerline Crossover	Go Long when transition from RSI < 50 to RSI > 50 Cover Long when transition from RSI > 50 to RSI < 50	Go Short when transition from RSI > 50 to RSI < 50 Cover Short when transition from RSI < 50 to RSI > 50
RSI Conservative Crossover	Go Long when transition from RSI < 50 to RSI > 50 Cover Long RSI > 75 (Overbought)	Go Short when transition from RSI > 50 to RSI < 50 Cover Short when RSI < 25 (Oversold)
RSI oversold/overbought	Go Long when RSI goes < 25 and then > 25 Cover Long RSI > 75 (Overbought)	Go Short when RSI goes > 75 and then < 75 Cover Short RSI < 25 (Overbought)
RSI Trend Reversal	Go Long when RSI trends up and Price trends down Cover Long Price trends up	Go Short when RSI trends down and Price trends up Cover short Price trends down
STOCH Overbought/Oversold	Go Long when Stoch goes < 20 and then > 20 Cover Long when Stoch > 80	Go Short when Stoch goes > 80 and then < 80 Cover Short when Stoch < 20
STOCH *k* Crossover	Go Long when $\%k > \%D$ Cover Long when $\%k < \%D$	Go Short when $\%k < \%D$ Cover Short when $\%k > \%D$
STOCH Trend Reversal	Go Long when Stoch trends up and Price trends down Cover Long Price trends up	Go Short when Stoch trends down and Price trends up Cover short Price trends down
ELDER Trend Reversal	(a) bear power < 0 but rising and (b) the previous bull power peak > the previous peak.	(a) bull power is > 0 but falling and (b) the previous bear power trough < the previous trough.
Bollinger Band Trade Rules	Go Long when Price crosses lower bollinger band	Go Short when Price crosses upper bollinger band

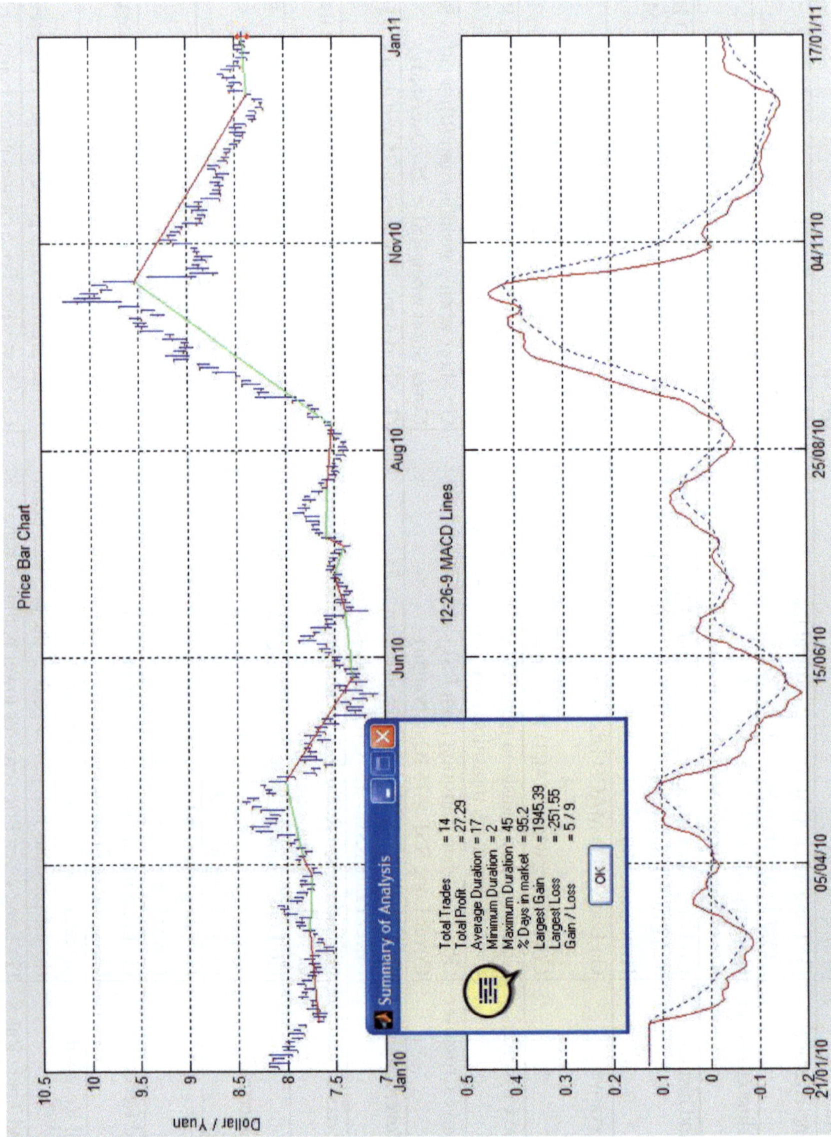

Fig. 7.19 Investment simulation result

6. *Best Stock Indicator Determination*
   The 'determine the best indicator' function on the main panel (see Figure 7.1), as its name suggests, can be used to determine the best indicator that one should use to trade a particular stock counter. Shown in Figure 7.20 is a sample evaluation result, which indicates that the best trading strategy is to use the MACD with the moving average crossover rule.

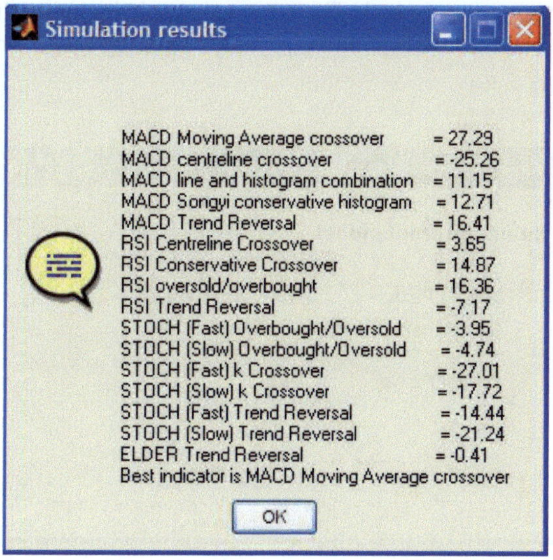

Fig. 7.20 Result of a best stock indicator determination

Lastly, we should note once again that all the trading rules and analysis tools and techniques presented above are just for reference and research purpose. The authors bear no responsibilities whatsoever to any gain or loss that one might generate from the stock market.

7.2.4 System Adaptation Framework

Several functions related to the system adaptation framework, i.e., the causality test and market turning period forecasting, have also been implemented in the toolkit. The causality test is mainly for advanced users who conduct research in financial engineering, whereas the market turning period forecasting function is useful for general users. Nonetheless, these functions are very time-consuming.

7.2.4.1 Causality Tests

Both the linear time-varying and nonlinear causality tests are implemented with an predetermined OE model. Users need to designate and input a related OE model (see Figure 7.21) and prepare the data of potential influential factors in an excel file (see Figure 7.22) in the case that the testing data cannot be downloaded online from the Internet. Since the oil price and the VXD are always considered as the key market influential factors and their data are readily available online, we display them explicitly on the causality test panel as depicted in Figure 7.22, which shows the test results (causality strengths and threshold values) of the internal residue and the influential factor. For the nonlinear causality test, users need to select the range of data to be further investigated. The returned result includes the values of CS and TVAL as well as the corresponding p values with lag length from 1 to 10 (see Figure 7.23).

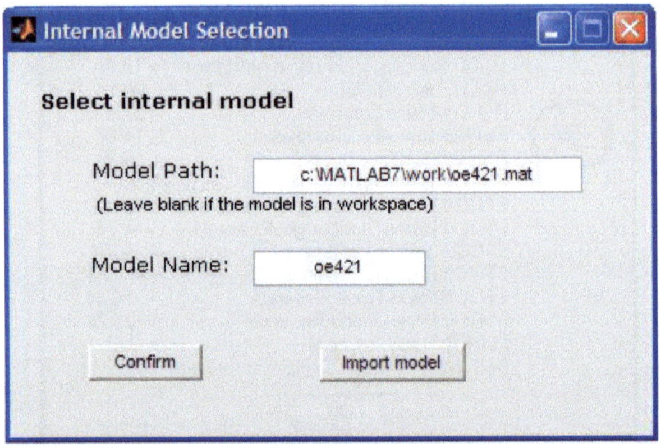

Fig. 7.21 OE model selection

7.2.4.2 Turning Period Forecasting and Confirmation

Another key feature of the T-TAS toolkit is the forecasting and confirmation of the major market turning periods. As studied in Chapter 6, our approach consists of two steps: (i) forecasting using the characteristic patterns associated with the frequency response of the internal residue and (ii) confirmation using the unstable points of the internal OE model. The default starting date of forecating is the beginning of the stock data under studied. Default parameters could also be adjusted. Shown in Figures 7.24 and 7.25 are, respectively, the sample results of the forecasting and confirmation of the major market turning periods for the Dow Jones Industrial Average Index from 1996 to 2011.

Finally, we note that we might update our toolkit from time to time. Once again, interested readers are referred to the T-TAS website managed and maintained by the authors at http://uav.ece.nus.edu.sg/~bmchen/ for the most up-to-date features of the toolkit.

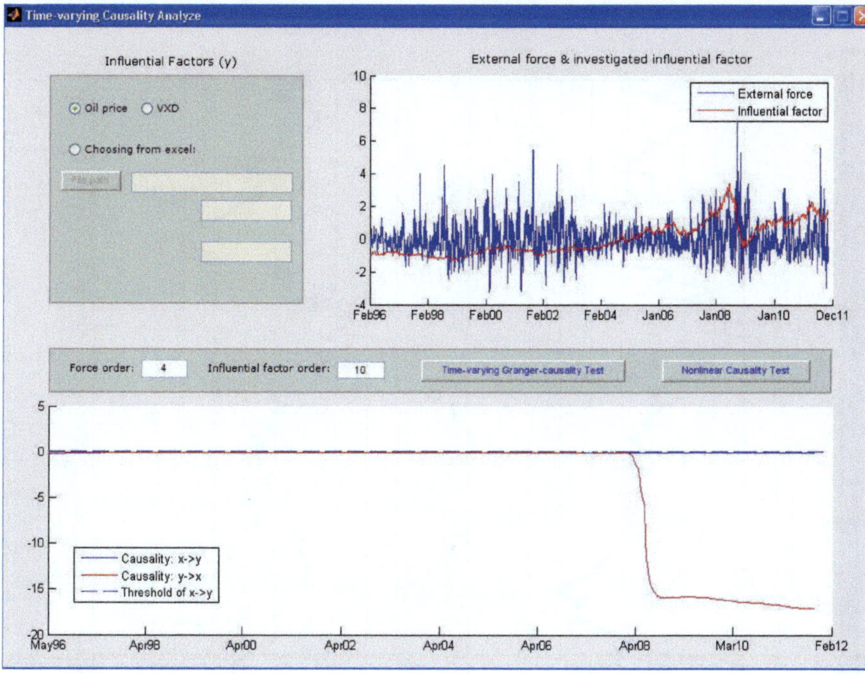

Fig. 7.22 Interface and result of the linear causality test

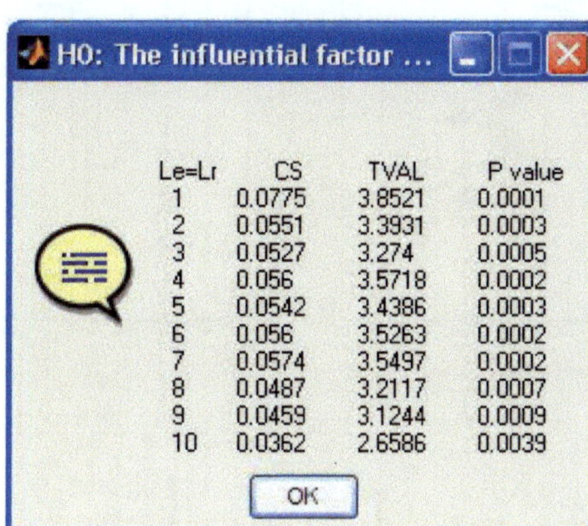

Le=Lr	CS	TVAL	P value
1	0.0775	3.8521	0.0001
2	0.0551	3.3931	0.0003
3	0.0527	3.274	0.0005
4	0.056	3.5718	0.0002
5	0.0542	3.4386	0.0003
6	0.056	3.5263	0.0002
7	0.0574	3.5497	0.0002
8	0.0487	3.2117	0.0007
9	0.0459	3.1244	0.0009
10	0.0362	2.6586	0.0039

Fig. 7.23 Result of the nonlinear causality test

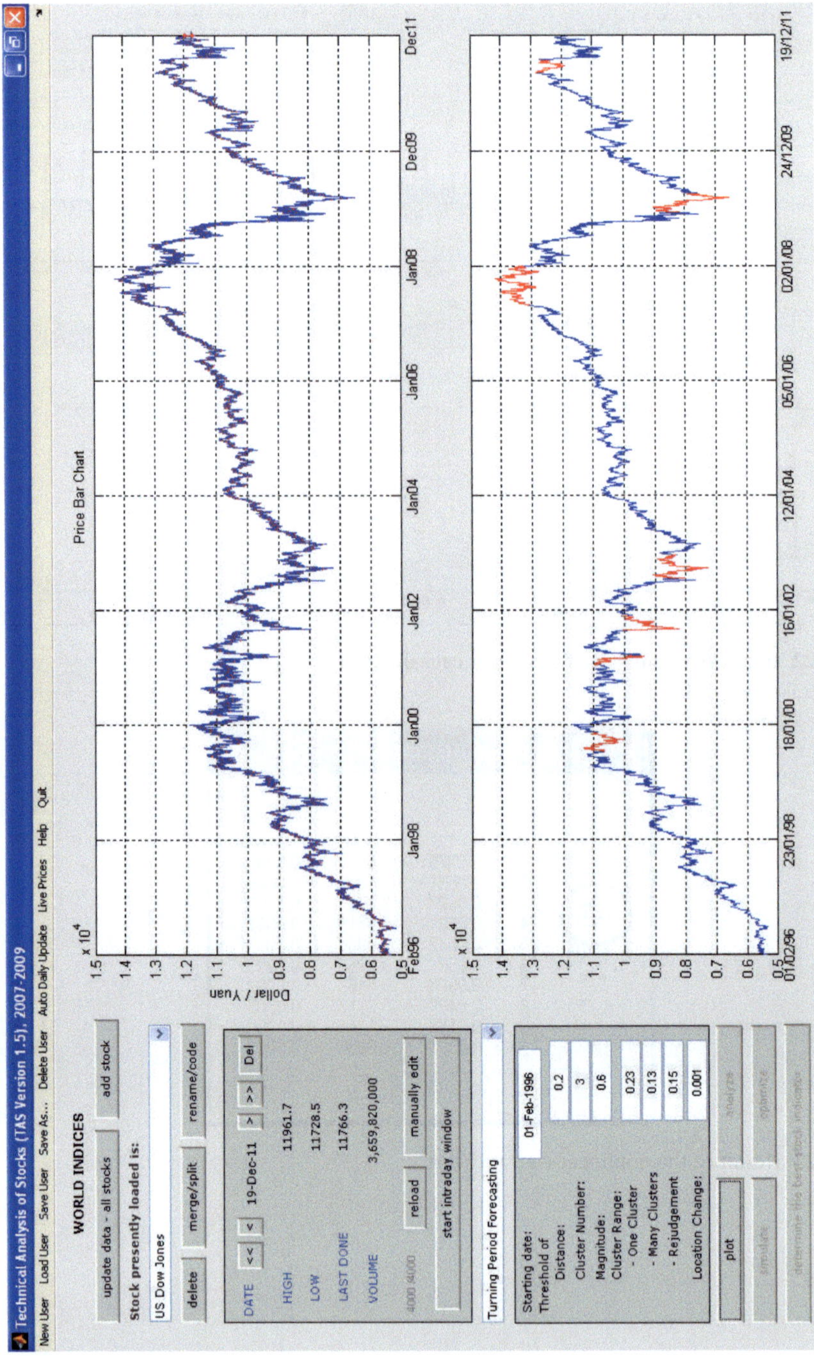

Fig. 7.24 Market turning period forecasting

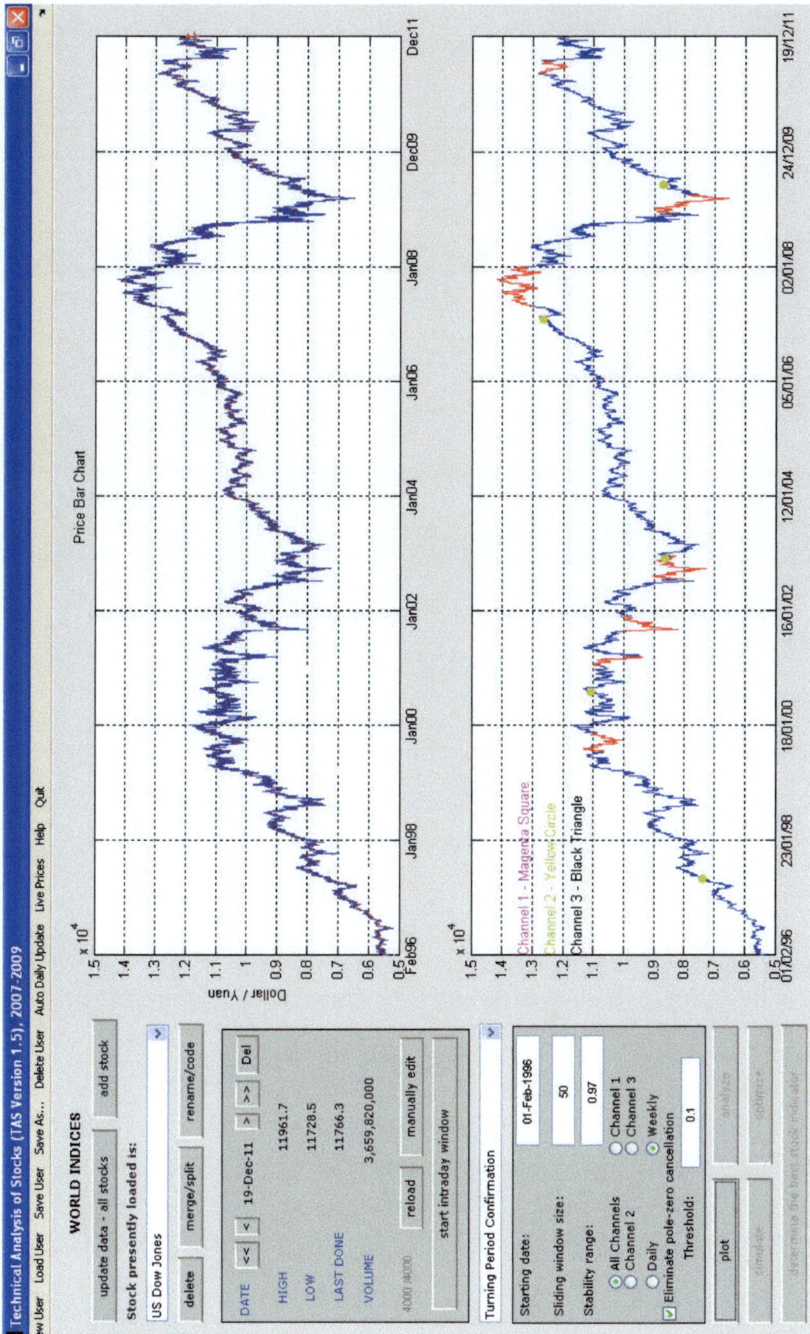

Fig. 7.25 Market turning period confirmation

Chapter 8
Future Research

We have documented in this monograph a system adaptation framework to systematically model the dynamics of the stock market. The framework consists of two parts: (i) an internal OE model, which is to capture the market slow dynamics, i.e., the market trends, and (ii) a time-varying adaptive filter, which is to capture the market fast dynamics corresponding to the influence of external forces. With appropriately selected input influential factors, OE model and adaptive filter, the system adaptation framework has shown to yield excellent prediction performance for the stock markets in the U.S., China, Singapore and Hong Kong. It has also been tested that the framework can be adopted to forecast the major turning periods in the stock market.

Even though we have laid out a fundamental structure for modeling financial systems using systems theory, the work, however, is by no means complete. As mentioned earlier, the results that we have obtained and documented in this monograph are very preliminary. More work is needed in order to enrich and mature the research along the line. We highlight in the following some observations that we have made. It is our belief that these topics are interesting and important and they are worth further investigation.

Structurally, the system adaptation framework proposed in this monograph has only two layers to, respectively, capture the fast and slow market dynamics. Due to the extremely high complexity and volatility of the stock market, a two-layer framework might not be sufficient enough to thoroughly identify all the key properties of the stock market. A more comprehensive structure might be needed to model a financial system (such as models for high frequency trading) that exhibits multiple time-scaled dynamical behaviors.

Within the system adaptation framework studied in this entire monograph, the research for a more efficient and effective internal model and adaptive filter is still an open problem. It is evident from the result obtained in Chapter 6 that the system stability of the internal model plays an interesting role in confirming the forecast of the major market turning period. Another research direction would thus be to further investigate the properties of the model identified using the very rich system theory techniques. For example, one might follow the system decomposition techniques

X. Zheng & B.M. Chen: *Stock Market Modeling and Forecasting*, LNCIS 442, pp. 149–150.
DOI: 10.1007/978-1-4471-5155-5_8

of [27] to further decompose the identified internal model into various subsystems, which, along with the interconnections that exist among them, would reveal more structural properties of the market.

It is evident that the selection of market influential factors directly affect the prediction result under the system adaptation framework. As mentioned earlier, how to incorporate more market information and how to systematically select appropriate influential factors that are suitable for the specific market system of interest could be an important direction for future research. For example, high frequency trading has attracted a lot of attention nowadays. How to identify influential factors for high frequency trading could be a very impactful area.

Lastly, but certainly not the least, as we have pointed out in Chapter 6, the rules that have adopted in forecasting the major market turning periods are very premature. They are neither rigorous nor perfect. We believe that it is possible to employ some appropriate filters, as those used in signal processing for engineering systems, to block out the unwanted components in the frequency response of the internal residue, under which a more rigorous procedure can then be developed to effectively identify the market turning periods. This could be another interesting topic for future research.

References

1. Aggarwal, R., Aggarwal, R.: Security return distributions and market structure: Evidence from the NYSE/AMEX and the NASDAQ markets. Journal of Financial Research 16(3), 209–220 (1993)
2. Albertos, P., Mareels, I.: Feedback and control for everyone. Springer, Berlin (2010)
3. Alles, L., Kling, J.: Regularities in the variation of skewness in stock index returns. Journal of Financial Research 17(3), 427–438 (1994)
4. Andrew, D.W.K., Ploberger, W.: Optimal tests when a nuisance parameter is present only under the alternative. Econometrica 62(6), 1383–1414 (1994)
5. Ariel, R.A.: A monthly effect in stock returns. Journal of Financial Economics 18(1), 161–174 (1987)
6. Badrinath, S.G., Chatterjee, S.: A data-analytic look at skewness and elongation in common-stock-return distributions. Journal of Business and Economic Statistics 9(2), 223–233 (1991)
7. Baek, E., Brock, W.: A general test for non-linear Granger causality: Bivariate model. In: Working papers. Iowa State University and University of Wisconsin, Madison (1992), `http://www.ssc.wisc.edu/~wbrock/BaekBrockGranger.pdf`
8. Bai, J., Perron, P.: Estimating and testing linear models with multiple structural changes. Econometrica 66(1), 47–78 (1998)
9. Bai, J., Perron, P.: Computation and analysis of multiple structural change models. Journal of Applied Econometrics 18(1), 1–22 (2003)
10. Baker, M., Wurgler, J.: Investor sentiment and the cross-section of stock returns. Journal of Finance 61(4), 1645–1680 (2006)
11. Baker, M., Wurgler, J.: Investor sentiment in the stock market. Journal of Economic Perspectives 21(2), 129–152 (2007)
12. Banathy, B.H.: A Taste of Systemics. In: ISSS The Primer Project (1997)
13. Beine, M., Capelle-Blancard, G., Raymond, H.: International nonlinear causality between stock markets. The European Journal of Finance 14(8), 663–686 (2008)
14. Bekaert, G., Erb, C.B., Harvey, C.R., Viskanta, T.E.: Distributional characteristics of emerging market returns and asset allocation. The Journal of Portfolio Management 24(2), 102–116 (1998)
15. Bernanke, B.S., Kuttner, K.N.: What explains the stock market's reaction to federal reserve policy? Journal of Finance 60(3), 1221–1257 (2005)
16. Binder, J.J., Merges, M.J.: Stock market volatility and economic factors. Review of Quantitative Finance and Accounting 17(1), 5–26 (2001)

17. Bachelier, L.: Théorie de la spéculation. Annales Scientifiques de lecole Normale Supérieure 17, 21–86 (1900)
18. Bachelier, L.: Théorie de la spéculation. Gauthier-Villars, Paris (1900)
19. Black, F., Scholes, M.: The pricing of options and corporate liabilities. Journal of Political Economy 81(3), 637–654 (1973)
20. Bodurtha Jr., J.N., Cho, D.C., Senbet, L.W.: Economic forces and the stock market: An international perspective. Global Finance Journal 1, 21–46 (1989)
21. Bouman, D., Jacobsen, B.: The Halloween indicator, "sell in and go away": Another puzzle. American Economic Review 92(5), 1618–1635 (2002)
22. Box, G.E.P., Jenkins, G.M.: Time series analysis: Forecasting and control. Holden-Day, San Francisco (1970)
23. Bressler, S.L., Seth, A.K.: Wiener-Granger causality: A well established methodology. NeuroImage 58(2), 323–329 (2010)
24. Cabrera, D., Colosi, L., Lobdell, C.: Systems thinking. Evaluation and Program Planning 31(3), 299–310 (2008)
25. Campbell, J.Y., Ammer, J.: What moves the stock and bond markets? A variance decomposition for long-term asset returns. Journal of Finance 48(1), 3–37 (1993)
26. Cao, X.-R., Wang, D.X.: How good is technique analysis?. A keynote speech at the 6th International Conference on Computer Science and Education, Singapore (2011)
27. Chen, B.M., Lin, Z., Shamash, Y.: Linear systems theory: A structural decomposition approach. Birkhäuser, Boston (2004)
28. Chen, H., Lobo, B.J., Wong, W.-K.: Links between the Indian, U.S. and Chinese stock markets. In: Working papers, no wp0602. National University of Singapore, Singapore (2006)
29. Chen, N.-F., Roll, R., Ross, S.A.: Economic forces and the stock market. Journal of Business 59(3), 383–403 (1986)
30. Chen, P.: A random walk or color chaos on the stock market? – Time-frequency analysis of S&P indexes. Studies in Nonlinear Dynamics & Econometrics 1(2), 87–103 (1996)
31. Chen, P.: Trends, shocks, persistent cycles in evolving economy: Business cycle measurement in time-frequency representation. In: Barnett, W.A., Kirman, A.P., Salmon, M. (eds.) Nonlinear Dynamics and Economics, ch. 13, pp. 307–331. Cambridge University Press, Cambridge (1996)
32. Chen, S.-H., Liao, C.-C.: Agent-based computational modeling of the stock price-volume relation. Information Sciences 170(1), 75–100 (2005)
33. Chen, S.-H., Yeh, C.-H.: On the emergent properties of artificial stock markets: The efficient market hypothesis and the rational expectations hypothesis. Journal of Economic Behavior & Organization 49(2), 217–239 (2002)
34. Chen, S.Y.: Development of a MATLA toolkit for technical analysis of stock markets. Bachelor of Engineering Thesis, Department of Electrical and Computer Engineering, National University of Singapore, Singapore (2007)
35. Chiarella, C., Gao, S.: Modelling the value of the S&P 500 – A system dynamics perspective. In: Working papers, (115). University of Technology, Sydney (2002)
36. Chow, G.C.: Tests of equality between sets of coefficients in two linear regressions. Econometrica 28(3), 591–605 (1960)
37. Chow, G.C., Lawler, C.C.: A time series analysis of the Shanghai and New York stock price indices. Annals of Economics and Finance 4(1), 17–35 (2003)
38. Dekker, A., Sen, K., Young, M.R.: Equity market linkages in the Asia Pacific region: A comparison of the orthogonalised and generalised VAR approaches. Global Finance Journal 12(1), 1–33 (2001)

39. Ding, J., Ding, F., Zhang, S.: Parameter identification of multi-input, single-output systems based on FIR models and least squares principle. Applied Mathematics and Computation 197(1), 297–305 (2008)
40. Ding, M., Bressler, S.L., Yang, W., Liang, H.: Short-window spectral analysis of cortical event-related potentials by adaptive multivariate autoregressive modelling: Data preprocessing, model validation, and variability assessment. Biological Cybernetics 83(1), 35–45 (2000)
41. Drakos, K.: Terrorism activity, investor sentiment, and stock returns. Review of Financial Economics 19(3), 128–135 (2010)
42. Dwyer Jr., G.P., Locke, P.R., Yu, W.: Index arbitrage and nonlinear dynamics between the S&P 500 futures and cash. Review of Financial Studies 9(1), 301–332 (1996)
43. Elder, A.: Trading for a living: Psychology, trading tactics, money management. John Wiley, New York (1993)
44. Engle, R.F.: Autoregressive conditional heteroscedasticity with estimates of variance of United Kingdom inflation. Econometrica 50(4), 987–1008 (1982)
45. Fabozzi, F.J., Modigliani, F.G., Jones, F.J.: Foundations of financial markets and institutions, 4th edn. Prentice Hall, New Jersey (2009)
46. Fama, E.F.: The behavior of stock market prices. Journal of Business 38(1), 34–105 (1965)
47. Fama, E.F.: Efficient capital markets: A review of theory and empirical work. Journal of Finance 25(2), 383–417 (1970)
48. Fama, E.F., French, K.R.: Permanent and temporary components of stock prices. Journal of Political Economy 96(2), 246–273 (1988)
49. Friedman, J., Shachmurove, Y.: Using vector autoregression models to analyze the behavior of the European community stock markets. In: CARESS working papers, (97-04). University of Pennsylvania, Philadelphia (1997)
50. Garefalakis, A.E., Dimitras, A., Koemtzopoulos, D., Spinthiropoulos, K.: Determinant factors of Hong Kong stock market. International Research Journal of Finance and Economics 62, 50–60 (2011)
51. Gerencsér, L., Mátyás, Z.: A behavioral stock market model. Mathematical Methods of Operations Research 67(1), 43–63 (2008)
52. Geweke, J.F.: Measurement of linear dependence and feedback between multiple time series. Journal of the American Statistical Association 77(378), 304–313 (1982)
53. Geweke, J.F.: Measures of conditional linear dependence and feedback between time series. Journal of the American Statistical Association 79(388), 907–915 (1984)
54. Golob, J.E., Bishop, D.G.: Do stock prices follow interest rates or inflation. In: Research Working Papers, (96-13). Federal Reserve Bank of Kansas City, Kansas (1996)
55. Google Finance, `http://www.google.com/finance` (cited 2013)
56. De Gooijera, J.G., Sivarajasingham, S.: Parametric and nonparametric Granger causality testing: Linkages between international stock markets. Physica A: Statistical Mechanics and its Applications 387(11), 2547–2560 (2008)
57. Graham, B., Dodd, D.: Security analysis, 1st edn. McGraw-Hill, New York (1934)
58. Granger, C.W.J.: Investigating causal relations by econometric models and cross-spectral methods. Econometrica 37(3), 424–438 (1969)
59. Granger, C.W.J.: Forecasting in business and economics, 2nd edn. Academic Press, San Diego (1989)
60. Guo, H.: Stock prices, firm size, and changes in the federal funds rate target. The Quarterly Review of Economics and Finance 44(4), 487–507 (2004)
61. Hanssens, D.M., Parsons, L.J., Schultz, R.L.: Market response models: Econometric and time series analysis. Kluwer Academic Publishers, Boston (1990)

62. Harvey, A.C.: Forecasting structural time series models and the Kalman filter. Cambridge University Press, Cambridge (1989)
63. Hesse, W., Möller, E., Arnold, M., Schack, B.: The use of time-variant EEG Granger causality for inspecting directed interdependencies of neural assemblies. Journal of Neuroscience Methods 124(1), 27–44 (2003)
64. Hiemstra, C., Jones, J.D.: Some information relating to the finite sample properties of the modified Baek and Brock nonlinear Granger causality test. In: Working papers. University of Strathclyde and Securities and Exchange Commission, Glasgow (1993)
65. Hiemstra, C., Jones, J.D.: Testing for linear and nonlinear Granger causality in the stock price-volume relation. Journal of Finance 49(5), 1639–1664 (1994)
66. Kim, K.-H.: Dollar exchange rate and the stock price: Evidence from multivariate cointegration and error correction model. Review of Financial Economics 12(3), 301–313 (2003)
67. Hol, E.M.J.H.: Empirical studies on volatility in international stock markets. Kluwer Academic Publishers, Boston (2003)
68. Hsieh, D.A.: Chaos and nonlinear dynamics: Application to financial markets. Journal of Finance 46(5), 1839–1877 (1991)
69. Huang, L., Wang, C., Qin, S.: A multiresolution wavelet based analysis of the Chinese stock market. In: Proceedings of the Third International Conference on Business Intelligence and Financial Engineering, Hong Kong, pp. 305–309 (August 2010)
70. IMF Survey Online, Asia's Importance Growing in Global Economy. In: Asia in the world economy. IMF Survey Magazine: Countries & Regions (2010), `http://www.imf.org/external/pubs/ft/survey/so/2010/car051210a.htm`
71. Ito, S., Tay, B.N.: The impacts of the oil shock on the Singaporean economy: Simulation results of a CGE model. Asian Economic Journal 8(1), 59–83 (1994)
72. Jang, G.-S., Lai, F., Jiang, B.-W., Chien, L.-H.: Intelligent stock trading decision support system with price trend prediction and reversal recognition using dual-module neural networks. Journal of Applied Intelligence 3, 225–248 (1993)
73. Kanas, A.: Nonlinearity in the stock price–dividend relation. Journal of International Money and Finance 24(4), 583–606 (2005)
74. Kaplanski, G., Levy, H.: Sentiment and stock prices: The case of aviation disasters. Journal of Financial Economics 95(2), 174–201 (2010)
75. Kim, K.J.: Financial time series forecasting using support vector machines. Neurocomputing 55(1-2), 307–319 (2003)
76. Kim, S., In, F.: The relationship between stock returns and inflation: New evidence from wavelet analysis. Journal of Empirical Finance 12(3), 435–444 (2005)
77. Kimoto, T., Asakawa, K., Yoda, M., Takeoka, M.: Stock market prediction system with modular neural networks. In: Proceedings of the International Joint Conference on Neural Networks, San Diego, vol. 1, pp. 1–6 (June 1990)
78. Koh, L.Y., Wu, Y.: Determinants of the Singapore Stock Exchange – A VAR Analysis. In: Kendall, J.D., Park, D. (eds.) East Asian Economic Issues, vol. 4, pp. 107–123. World Scientific, Singapore (1998)
79. Kothari, S.P., Lewellen, J., Warner, J.B.: Stock returns, aggregate earnings surprises, and behavioral finance. Journal of Financial Economics 79(3), 537–568 (2006)
80. Kuttner, K.N.: Monetary policy surprises and interest rates: Evidence from the Fed funds futures market. Journal of Monetary Economics 47(3), 523–544 (2001)
81. Lagarias, J.C., Reeds, J.A., Wright, M.H., Wright, P.E.: Convergence properties of the nelder-mead simplex method in low dimensions. SIAM Journal on Optimization 9(1), 112–147 (1998)

82. Laurencec, M., Cai, F., Qian, S.: Weak-form efficiency and causality tests in Chinese stock markets. Multinational Finance Journal 1(4), 291–307 (1997)
83. Le, T-H., Chang, Y.: The impact of oil price fluctuations on stock markets in developed and emerging economies. In: Economic Growth Centre Working Paper Series, (1103). Nanyang Technological University, Singapore (2011)
84. LeBaron, B.: Evolution and time horizons in an agent-based stock market. Macroeconomic Dynamics 5(2), 225–254 (2001)
85. LeBaron, B., Arthur, W.B., Palmer, R.: Time series properties of an artificial stock market. Journal of Economic Dynamics and Control 23(9-10), 1487–1516 (1999)
86. Lee, W.Y., Jiang, C.X., Indro, D.C.: Stock market volatility, excess returns, and the role of investor sentiment. Journal of Banking & Finance 26(12), 2277–2299 (2002)
87. LeRoy, S.F.: Risk aversion and the martingale property of stock prices. International Economic Review 14(2), 436–446 (1973)
88. Ljung, L.: System identification: Theory for the user, 2nd edn. Prentice Hall, New Jersey (1999)
89. Lo, A.W.: The adaptive markets hypothesis: Market efficiency from an evolutionary perspective. Journal of Portfolio Management 30th Anniversary Issue, 15–29 (2004)
90. Lo, A.W., MacKinlay, A.C.: A non-random walk down Wall Street. Princeton University Press, New Jersey (1999)
91. Lo, M.C., Piger, J.: Is the response of output to monetary policy asymmetric? Evidence from a regime-switching coefficient model. Journal of Money, Credit, and Banking 37(5), 865–886 (2005)
92. Lucas, R.: Asset prices in an exchange economy. Econometrica 46(6), 1429–1445 (1978)
93. Malkiel, B.G.: A random walk down wall street. W. W. Norton & Company, New York (1973)
94. Financial & Special Studies Division, Financial market integration in Singapore: The narrow and the broad views. In: Occasional Papers, (20). Economics Department, Monetary Authority of Singapore, Singapore (2000)
95. Maysami, R.C., Koh, T.S.: A vector error correction model of the Singapore stock market. International Review of Economics & Finance 9(1), 79–96 (2000)
96. McKay, P.A.: Oil lifts Dow to 10062.94 in late rally. Wall Street Journal, 1 (October 16, 2009)
97. Michael, D.I.: Interaction between economics and systems theory. In: Optimization Techniques: Lecture Notes in Control and Information Sciences, vol. 22, pp. 124–130. Springer, Berlin (1980)
98. Mookerjee, R., Yu, Q.: Macroeconomic variables and stock prices in a small open economy: The case of Singapore. Pacific-Basin Finance Journal 5(3), 377–388 (1997)
99. Motiwalla, L., Wahab, M.: Predictable variation and profitable trading of US equities: A trading simulation using neural networks. Computers and Operations Research 27(11-12), 1111–1129 (2000)
100. Nieh, C.-C., Yau, H.-Y.: The impact of Renminbi appreciation on stock prices in China. Emerging Markets Finance and Trade 46(1), 16–26 (2010)
101. Niemira, M.P.: Forecasting turning points in the stock market cycle and asset allocation implications. In: Klein, P.A. (ed.) Analyzing Modern Business Cycles: Essays Honoring Geoffrey H. Moore, pp. 109–127. M. E. Sharpe, New York (1990)
102. Orrell, D., McSharry, P.E.: System economics: Overcoming the pitfalls of forecasting models via a multidisciplinary approach. International Journal of Forecasting 25(4), 734–743 (2009)

103. Peters, E.E.: Chaos and order in the capital markets: A new view of cycles, prices, and market volatility. John Wiley & Sons, New York (1991)
104. Peters, E.E.: Fractal market analysis: Applying chaos theory to investment and economics. John Wiley & Sons, New York (1994)
105. Phylaktis, K., Ravazzolo, F.: Stock prices and exchange rate dynamics. Journal of International Money and Finance 24(7), 1031–1053 (2005)
106. Poggio, T., Lo, A.W., LeBaron, B.D., Chan, N.T.: Agent-based models of financial markets: A comparison with experimental markets. In: MIT Sloan Working Papers, (4195–01). MIT Sloan School of Management, Cambridge (2001)
107. Psaradakis, Z., Ravn, M.O., Sola, M.: Markov switching causality and the moneyCoutput relationship. Journal of Applied Econometrics 20(5), 665–683 (2005)
108. Rapach, D.E., Wohar, M.E., Rangvid, J.: Macro variables and international stock return predictability. International Journal of Forecasting 21(1), 137–166 (2005)
109. Roberts, H.: Statistical versus clinical prediction of the stock market. Center for Research in Security Prices, University of Chicago, Chicago (1967) (unpublished manuscript)
110. Roebroeck, A., Formisano, E., Goebel, R.: Mapping directed influence over the brain using Granger causality and fMRI. Neuroimage 25(1), 230–242 (2005)
111. Samuelson, P.A.: Interactions between the multiplier analysis and the principle of acceleration. The Review of Economics and Statistics 21(2), 75–78 (1939)
112. Samuelson, P.A.: Proof that properly anticipated prices fluctuate randomly. Industrial Management Review 6, 41–49 (1965)
113. Sariannidis, N., Drimbetas, E., Konteos, G.: Impact of international volatility, the Euro, and derivatives on a small stock market. International Review of Applied Economic Research 1, 1–21 (2006)
114. Turhan-Sayan, G., Sayan, S.: Use of time-frequency representations in the analysis of stock market data. In: Kontoghiorghes, E.J., Rustem, B., Siokos, S. (eds.) Computational Methods in Decision-Making, Economics and Finance, ch. 22, pp. 429–453. Kluwer Academic Publishers, Dordrecht (2002)
115. Schinasi, G.J.: Defining Financial Stability. In: IMF Working Papers, (04/187). International Monetary Fund (2004)
116. Sharpe, W.F.: Capital asset prices: A theory of market equilibrium under conditions of risk. Journal of Finance 19(3), 425–442 (1964)
117. Sharpe, W.F., Alexander, G.J., Bailey, J.V.: Investments, 6th edn. Prentice Hall, New Jersey (1998)
118. Shleifer, A.: Inefficient markets: An introduction to behavioral finance. Oxford University Press, New York (2000)
119. Silvapulle, P., Choi, J.-S.: Testing for linear and nonlinear Granger causality in the stock price-volume relation: Korean evidence. The Quarterly Review of Economics and Finance 39(1), 59–76 (1999)
120. Sims, C.A.: Money, income and causality. American Economic Review 62(4), 540–552 (1972)
121. Steel, R.G.D., Torrie, J.H.: Principles and procedures of statistics: A biometrical approach. McGraw-Hill, New York (1960)
122. Sun, W.: A comprehensive approach to predicting market bottoms. Bachelor of Engineering Thesis, Department of Electrical & Computer Engineering, National University of Singapore, Singapore (2009)
123. Taylor, C.J., Pedregal, D.J., Young, P.C., Tych, W.: Environmental time series analysis and forecasting with the Captain toolbox. Environmental Modelling and Software 22(6), 797–814 (2007), `http://www.es.lancs.ac.uk/cres/captain`

124. Thoma, M.A.: Subsample instability and asymmetries in Money-Income causality. Journal of Econometrics 64(1-2), 279–306 (1994)
125. Thornton, D.L.: Tests of the market's reaction to federal funds rate target changes. Review, Federal Reserve Bank of St. Louis 1998, 25–36 (1998)
126. Tian, G.G., Ma, S.: The relationship between stock returns and the foreign exchange rate: The ARDL approach. Journal of the Asia Pacific Economy 15(4), 490–508 (2010)
127. Turner, A.L., Weigel, E.J.: An analysis of stock market volatility. In: Russell Research Commentaries. Frank Russell Company, Tacoma (1990)
128. Wigner, E.P.: On the quantum correction for thermodynamic equilibrium. Physical Review 40(5), 749–759 (1932)
129. Wong, J., Fung, L.: Liquidity of the Hong Kong stock market since the Asian financial crisis. In: Working Papers. Hong Kong Monetary Authority, Hong Kong (2001)
130. Wu, Y.: Exchange rates, stock prices, and money markets: Evidence from Singapore. Journal of Asian Economics 12(3), 445–458 (2001)
131. Yahoo Finance, `http://finance.yahoo.com/` (cited 2013)
132. Yang, H.Q.: Margin variations in support vector regression for the stock market prediction. Master thesis, Chinese University of Hong Kong, Hong Kong (2003)
133. Yang, Y.: The relationship between stock and foreign exchange markets: Evidence from periods of exchange-rate-regime shifts. Master Thesis, Department of Economics and Finance, City University of Hong Kong, Hong Kong (2009)
134. Yao, F., Ying, Y., Zhang, L.: Analysis of stock market information - A new financial engineering approach. In: Proceedings of IEEE International Conference on Information and Automation, Harbin, pp. 430–435 (June 2010)
135. Yao, S., Luo, D., Stephen, M.: Shanghai stock exchange composite index and bank stock prices in China: A causality analysis. In: Research Paper Series, (08/25). University of Nottingham, Nottingham (2008)
136. Young, P.C.: Time-variable parameter and trend estimation in non-stationary economic time series. Journal of Forecasting 13(2), 179–210 (1994)
137. Young, P.C.: Stochastic, dynamic modeling and signal processing, time variable and state dependent parameter estimation. In: Fitzgerald, W.J., Smith, R.L., Walden, A.T., Young, P.C. (eds.) Nonlinear and Nonstationary Signal Processing, pp. 74–114. Cambridge University Press, Cambridge (2000)
138. Yu, I.-W., Tam, C.-S.: Measuring market sentiment in Hong Kong's stock market – recent development of the Hang Seng China Enterprise Index. In: Working Papers, (04/2007). Hong Kong Monetary Authority, Hong Kong (2007)
139. Yu, Q.: Stock prices and exchange rates: Experience in leading East Asian financial centres: Tokyo, Hong Kong and Singapore. Singapore Economic Review 41(1), 47–56 (1997)
140. Zheng, X., Chen, B.M.: Identification of stock market forces in the system adaptation framework. In: Proceedings of the 9th IEEE International Conference on Control and Automation, Santiago, pp. 340–347 (December 2011)
141. Zheng, X., Chen, B.M.: Modeling and forecasting of stock markets under a system adaptation framework. Journal of Systems Science and Complexity 25(4), 641–674 (2012)
142. Zheng, Y., Wong, W.-K.: Mean, volatility spillover and time-varying conditional dependence in Chinese stock markets. In: RMI Working Papers, (07/02). National University of Singapore, Singapore (2007)

Index

MIX
Papier aus verantwortungsvollen Quellen
Paper from responsible sources
FSC® C105338

If you have any concerns about our products,
you can contact us on
ProductSafety@springernature.com

In case Publisher is established outside the EU,
the EU authorized representative is:
Springer Nature Customer Service Center GmbH
Europaplatz 3, 69115 Heidelberg, Germany

Printed by Libri Plureos GmbH
in Hamburg, Germany